THE
WEATHER
IDENTIFICATION HANDBOOK

STORM DUNLOP

THE LYONS PRESS
GUILFORD, CONNECTICUT
AN IMPRINT OF THE GLOBE PEQUOT PRESS

The Lyons Press is an imprint of The Globe Pequot Press

10 9 8 7 6 5 4 3 2 1

Printed in Singapore

Diagrams by chrisom.com
Color reproductions by Colourscan, Singapore
Edited and designed by D & N Publishing, Baydon, Wiltshire
Manufactured by Imago

ISBN 1-58574-857-9

Library of Congress Cataloging-in-Publication Data is available on file.

CONTENTS

How to use this book

This book concentrates on identification of the many different types of phenomena that may be seen, but also gives brief details of the weather that may be expected. The various tables provide a guide to where full details are given. The major cloud types may be identified using the table on p. 9, and the descriptions of the many different forms may be located using the tables (pp. 10–12). A section on the formation of clouds (beginning p. 86), describes some more general, distinctive forms often visible.

Optical phenomena may be identified with the help of the lists given (pp. 102–103), while sky colors and shadow effects are dealt with a separate section (pp. 128–141), which is followed by a short section (pp. 142–145) on visibility. The various types of precipitation (dew, rain, snow, hail, etc.) are covered next (pp. 146–155).

Information about winds (pp. 156–163) is followed by a description of various forms of severe weather (pp. 164–175), and a brief summary of the overall global circulation and how this relates to weather charts (pp. 176–183). A table (pp. 184–185) shows, in highly simplified form, how current cloud types relate to possible forthcoming weather.

Readers are strongly recommended to begin by learning the ten major cloud types, together with some of the more distinctive subsidiary forms, all of which are covered in the first major section (pp. 10–79). It will also help to follow the simple tips for observing and photographing the sky that are given on p. 6. A glossary begins on p. 188.

OBSERVING THE SKY

Some types of cloud and other weather phenomena may be recognized at a glance, but others may require more time or practice. Generally, many problems of identification may be resolved by taking a bit of time to watch the sky.

Because clouds at different levels generally move at different speeds—or even in different directions—watching the sky for a while may show that what at first sight appears to be a chaotic sky, full of different types of clouds, actually consists of specific types at particular altitudes.

It is useful to have a rough method of estimating the angular size of particular phenomena. In doubtful cases, for example, the size of the cloud elements helps to determine whether a particular cloud should be classed as stratocumulus, altocumulus, or cirrocumulus. The simplest method of estimating sizes is to use one's hand, held at arm's length as shown in the diagram. This is surprisingly accurate, and certainly suffices for general identification purposes.

It may seem obvious, but few people think to use binoculars to examine clouds. Yet this is sometimes a simple way of determining the precise nature of certain clouds and watching one form change into another. The magnified view may also help to reveal the motion of air around or within the cloud. If using binoculars, however, do remember not to look at the Sun with them, because eye damage could result.

It is often difficult to see details of clouds or optical phenomena (especially near the Sun) because the sky and clouds appear so bright. Blocking out the Sun with a hand or some suitable object will often suffice, but there are various other ways of improving the situation:

- VIEW A REFLECTION OF THE SKY
- USE SUNGLASSES
- USE A POLARIZER

Viewing the sky as reflected in a pool of water or glass will help. With water, a pool that is shaded, deep, and has a dark bottom

Some effects are very rare. Here the third arc, rising vertically from the base of the primary rainbow, is a reflected-light bow, only three photographs of which are known

is most effective. With glass, a dark glass is best, but the tinted glass used in many modern buildings is quite useful. Although any sunglasses will help, the mirror type are by far the best. Two pieces of polarizing material may be turned relative to one another to produce any degree of darkening required.

It is easy to estimate angles on the sky by holding one's hand at arm's length

A single piece of polarizing material, which may be bought as a thin plastic sheet, is useful on its own. The sky background itself is quite strongly polarized in certain directions (particularly 90° from the Sun). Turning a polarizer will accentuate (or diminish) the contrast of clouds against the sky, the visibility of different parts of clouds, and the colors of certain optical phenomena, such as rainbows and haloes.

Photographing the sky and clouds

Taking photographs of the sky, clouds, or optical phenomena is relatively easy. Most automatic-exposure systems will produce acceptable results, particularly if you are prepared for any foreground to be slightly dark. Although photographs may be obtained with any form of camera, including digital ones, a single-lens reflex with interchangeable lenses remains the best choice. A non-standard wide-angle lens may be required to include certain optical phenomena: a 28-mm lens will be needed to capture the full extent of a secondary rainbow on 35-mm film, for example.

A lens hood is essential and it is good practice to use one at all times, because it is only too easy to point the camera close to the Sun when photographing clouds, and find the image spoiled by flare. A skylight filter will protect the lens and will also help to reduce excessive blue-violet coloration if photographing at high altitude. A polarizer is highly desirable, and you should ensure that you have the correct type for your particular camera. Some require what are known as circular polarizers, and will not give correct exposures if a linear polarizer is used instead.

When photographing any unusual phenomena, ensure that you make a note of the focal length of the lens used, and particularly its setting if you use a zoom lens. Only with this information is it possible to calculate the angular size or position of a phenomenon on the final image, which may be of considerable scientific interest. Naturally, the date and time should always be recorded.

Cloud classification was first introduced in 1802 by Luke Howard, a British pharmacist, who later became an industrial chemist. He used a system that is very similar to the way in which plants and animals are classified, with names that are derived from Latin.

Howard's system has been kept to the present day, although considerably modified, and clouds are divided into genera, species, and varieties, together with what are known as accessory clouds and supplementary types. This may seem complicated, but don't let it worry you; there are just ten types that you need to recognize initially—far less than the classes of birds or flowers, for instance. Once you are able to identify these ten, you will find that the most common sub-types and variations will become relatively easy to recognize.

The overall scheme is able to describe clouds in great detail, but it is worth remembering that these descriptions are based solely on the clouds' appearance. The actual way in which clouds have formed, and their subsequent changes, may be of greater significance for understanding the changes in the weather that are occurring, or that may take place shortly. We will discuss these points later, when describing both the formation of clouds and individual weather situations.

The various names have standardized, two- or three-letter abbreviations, which are extensively used when writing down observations (and often when describing clouds) rather than the full, and sometimes slightly cumbersome identifications. The full classification scheme is:

CLOUD GENUS (PL. GENERA)

Ten basic cloud types, covering broad, overall characteristics: two-letter abbreviations.

GENUS	ABBR.	DESCRIPTION	PAGE
altocumulus	Ac	Heaps or rolls of cloud, showing distinct shading, and with clear gaps between them, in a layer at middle levels	46
altostratus	As	Sheet of featureless, white or grey cloud at middle levels	40
cirrocumulus	Cc	Tiny heaps of cloud with no shading, with clear gaps, in a layer at high levels	64
cirrostratus	Cs	Essentially featureless sheet of thin cloud at high levels	60
cirrus	Ci	Fibrous wisps of cloud at high levels	54
cumulonimbus	Cb	Large towering cloud extending to great heights, with ragged base and heavy precipitation	68
cumulus	Cu	Rounded heaps of cloud at low levels	18
nimbostratus	Ns	Dark grey cloud at middle levels, frequently extending down towards surface, and giving prolonged precipitation	44
stratocumulus	Sc	Heaps or rolls of cloud, with distinct gaps and heavy shading at low levels	32
stratus	St	Essentially featureless, grey layer cloud at low level	26

(OPPOSITE PAGE) A sheet of stratus cloud being held back by the hills of the Serra Cavallera in Catalonia

Cumulus mediocris above Castlerigg prehistoric stone circle near Keswick, Cumbria

CLOUD SPECIES

Fourteen terms to describe cloud shape and structure; three-letter abbreviations.

SPECIES	ABBR.	DESCRIPTION	GENERA	PAGE
calvus	cal	Tops of rising cells lose their hard appearance and become smooth	Cb	70
capillatus	cap	Tops of rising cells become distinctly fibrous or striated; obvious cirrus may appear	Cb	70
castellanus	cas	Distinct turrets rising from an extended base or line of cloud	Sc, Ac, Cc, Ci	36, 49
congestus	con	Great vertical extent; obviously growing vigorously, with hard, 'cauliflower-like' tops	Cu	24
fibratus	fib	Fibrous appearance, normally straight or uniformly curved; no distinct hooks	Ci, Cs	56
floccus	flo	Individual tufts of cloud, with ragged bases, sometimes with distinct virga	Ac, Cc, Ci	50
fractus	fra	Broken cloud with ragged edges and base	Cu, St	22, 30
humilis	hum	Cloud of restricted vertical extent; length much greater than height	Cu	23
lenticularis	len	Lens- or almond-shaped clouds, stationary in the sky	Sc, Ac, Cc	50
mediocris	med	Cloud of moderate vertical extent, growing upwards	Cu	24
nebulosus	neb	Featureless sheet of cloud, with no structure	St, Cs	30
spissatus	spi	Dense cloud, appearing grey when viewed towards the Sun	Ci	57
stratiformis	str	Cloud in an extensive sheet or layer	Sc, Ac, Cc	36
uncinus	unc	Distinctly hooked, often without a visible generating head	Ci	57

CLOUD VARIETIES

Nine terms that describe cloud transparency and the arrangement of cloud elements: two-letter abbreviations. Any given cloud may exhibit the characteristics of more than one variety—often several may be present simultaneously.

VARIETY	ABBR.	DESCRIPTION	GENERA	PAGE
duplicatus	du	Two or more layers	Sc, Ac, Ac, Cc, Cs	37
intortus	in	Tangled or irregularly curved	Ci	58
lacunosus	la	Thin cloud with regularly spaced holes, appearing like a net	Ac, Cc, Sc	52
opacus	op	Thick cloud that completely hides Sun or Moon	St, Sc, Ac, As	31
perlucidus	pe	Extensive layer with gaps, through which blue sky, the Sun or Moon are visible	Sc, Ac	38
radiatus	ra	Appearing to radiate from one point in the sky	Cu, Sc, Ac, As, Ci	25, 53
translucidus	tr	Translucent cloud, through which the position of the Sun or Moon is readily visible	St, Sc, Ac, As	52
undulatus	un	Layer or patch of cloud with distinct undulations	St, Sc, Ac, As, Cc, Cs	31, 39
vertebratus	ve	Lines of cloud looking like ribs, vertebrae or fish bones	Ci	58

A layer of cirrostratus with billows (cirrostratus undulatus)

ACCESSORY CLOUDS

Three forms that occur only in conjunction with one of the 10 main genera: three-letter abbreviations.

NAME	ABBR.	DESCRIPTION	GENERA	PAGE
pannus	pan	Ragged shreds of cloud beneath main cloud mass	Cu, Cb, As, Ns	72
pileus	pil	Hood or cap of cloud above rising cell	Cu, Cb	72
velum	vel	Thin, extensive sheet of cloud, through which the most vigorous cells may penetrate	Cu, Cb	73

SUPPLEMENTARY FEATURES

Six particular forms (some common, others quite rare) that particular genera or species may adopt: three-letter abbreviations.

FEATURE	ABBR.	DESCRIPTION	GENERA	PAGE
arcus	arc	Arch or roll of cloud	Cb, Cu	74
incus	inc	Anvil cloud	Cb	75
mamma	mam	Bulges or pouches beneath higher cloud	Cb, Ci, Cc, Ac, As, Sc	76
praecipitatio	pre	Precipitation that reaches the surface	Cb, Cu, Ns	77
tuba	tub	Funnel cloud of any type	Cb, Cu	79
virga	vir	Fallstreaks: trails of precipitation that do not reach the surface	Ac, As, Cc, Cb, Cu, Ns, Sc, (Ci)	78

Two additional terms are used occasionally. These are the suffixes '-genitus' and '-mutatus'—abbreviated 'gen' and 'mut', respectively—which are added to a particular cloud genus to indicate the type from which a currently observed cloud has been derived. The first implies that a considerable amount of the parent cloud is still present; the second that essentially all the parent cloud has been altered. For instance, 'altocumulus cumulogenitus' (Ac cugen) indicates that the dominant cloud is altocumulus which has been formed from cumulus, substantial amounts of which persist. In the second case, 'stratus stratocumulomutatus' (St scmut) indicates the presence of stratus, originally derived from stratocumulus cloud.

Three of Howard's original terms for cloud types are still used unchanged, mainly because they accurately describe the three main forms that clouds may take. They are thus useful for dividing all the different

clouds into just three broad categories. Technically, meteorologists divide clouds into just the first two of the following groups, but the third appears to be a useful addition for initial identification purposes.

Massive cumuliform clouds building up over Pocahontas County, West Virginia

    CUMULUS [LATIN: 'HEAP'] — HEAP OR CUMULIFORM CLOUDS

    STRATUS [LATIN: 'LAYER'] — LAYER OR STRATIFORM CLOUDS

    CIRRUS [LATIN: 'CURL', 'TUFT', 'WISP'] — HAIR-LIKE OR CIRRIFORM CLOUDS

As we shall see later, this division is actually related both to the way in which the clouds form and to their composition. Even this simple division tells us something about the processes that are occurring in the

Stratus clouds
over South Georgia
caused by uplift over
the mountainous
island

atmosphere, and thus about the weather and the way in which it may develop. A quick glance at any cloud will enable you to place it in one of these broad groups.

Cumuliform clouds

 CUMULUS

CUMULONIMBUS

STRATOCUMULUS

ALTOCUMULUS

CIRROCUMULUS

All these types display more-or-less pronounced rounded heads or turrets on their upper surface. This is a sign that—as will be explained later— instability is present, and convection is occurring within the cloud.

Stratiform clouds

 STRATUS

NIMBOSTRATUS

ALTOSTRATUS

CIRROSTRATUS

In contrast to the cumuliform clouds, these layer clouds are an indication of atmospheric stability. They generally have smooth upper surfaces. Although the top of the layer may be impossible to see when such clouds blanket a large part of the sky, if there are breaks it is often possible to see that there are none of the rounded heads that are characteristic of the cumuliform types.

STRATOCUMULUS
ALTOCUMULUS
CIRROCUMULUS

SEE ALSO

convection (p. 86)

instability (p. 187)

stability (p. 188)

supercooling (p. 150)

Although these three types frequently occur in marked layers, their features indicate that shallow convection is occurring within them. They thus occupy an intermediate position between the cumuliform and stratiform groups.

Cirriform clouds

CIRRUS
CIRROCUMULUS
CIRROSTRATUS

These three types consist mainly of ice crystals, although cirrocumulus also contains supercooled water droplets. Cirrus and cirrostratus often show marked striations, caused by trails of ice crystals that appear thread- or hair-like.

As we see from these lists, some of the ten main cloud types appear in more than one group, because they have a combination of characteristics—cirrocumulus in all three. We will discuss the reasons for this later.

Typical cirriform cloud: cirrus fibratus with some cirrus floccus

CLOUD TYPES

The ten main cloud types (or genera, to use the official Latin term) are normally arranged by height, and this is the classification used in the official descriptions issued by the World Meteorological Organization. Cloud heights are often difficult to estimate, so this arrangement might seem perverse, but it is based on sound meteorological reasons. Luckily, the different cloud types are normally fairly easy to identify, so an accurate knowledge of a cloud's height is not necessary. In fact, the opposite holds true: from the cloud type it is possible to make a reasonable estimate of the height.

Again there are three broad categories: high, medium, and low clouds. In all cases, the height is taken to be that of the base of the cloud. The international aviation industry quotes altitudes in feet. Cloud heights vary considerably depending on one's latitude and the time of year, so the figures given here should be taken as approximate. We will discuss later some of the reasons for the variations and the ranges of height.

As with many aspects of cloud study, these divisions are somewhat artificial, because clouds may have a considerable vertical extent, and thus occur at more than one level. This is always the case with cumulonimbus, but also frequently occurs with certain cumulus clouds. Nimbostratus, too, may extend from the middle level down towards the surface, where it gives rise to heavy rain.

Similarly, clouds often occur that show one set of characteristics in one part of the sky, and different features in another. A sheet of altostratus may grade into altocumulus, for example. Such changes often also occur with time. In addition, of course, several different types of cloud may be present at any one time. The ten basic types do, however, form a useful starting point for identifying clouds. Although normally discussed in descending order of altitude, the order is reversed here, because lower clouds will be more familiar to most readers. Some cloud types (most notably the stratiform ones) may exist simultaneously at two or more heights, that is, as the variety known as duplicatus. When the lower layer is extensive, it may be difficult to decide whether the higher layer is truly the same or belongs to a different class.

HIGH CLOUDS BASES AT, OR ABOVE APPROX. 20,000 FT.

TROPICS 20,000 – 60,000 FT.　　MIDDLE LATITUDES 16,500 – 43,000 FT.　　HIGH LATITUDES 10,000 – 26,500 FT.

CIRRUS　　　　　　CIRROSTRATUS　　　　　　CIRROCUMULUS

MEDIUM-LEVEL CLOUDS BASES BETWEEN APPROX. 6,500–20,000 FT.

TROPICS 6,500 – 26,500 FT.　　MIDDLE LATITUDES 6,500 – 23,000 FT.　　HIGH LATITUDES 6,500 – 13,000 FT.

ALTOCUMULUS　　　　　　ALTOSTRATUS　　　　　　NIMBOSTRATUS

LOW CLOUDS BASES BELOW APPROX. 6,500 FT.

TROPICS 0 – 6,500 FT.　　MIDDLE LATITUDES 0 – 6,500 FT.　　HIGH LATITUDES 0,000 – 6,500 FT.

CUMULUS　　　　　　STRATUS　　　　　　STRATOCUMULUS

CLOUDS THAT EXTEND THROUGH MORE THAN ONE LEVEL

CUMULONIMBUS

CUMULUS (Cu)

Cumulus clouds are well-known to everyone. They are white heaps of cloud, usually with fairly distinct, smooth outlines and a flat, darker base. They occur so frequently in fine weather, floating in a blue sky, that they are commonly known as 'fair-weather clouds' (or 'fair-weather cumulus').

CHARACTERISTIC FEATURES

- Heap clouds, generally with a rounded, white top and darker base. One species (fractus) always appears ragged.
- Generally no precipitation, but one species (congestus) sometimes produces rain.

Cumulus clouds usually build up gradually during the day. Initially, they may be small tufts of cloud, often fairly ragged in appearance, that do not persist for very long. Gradually, as the day goes by, the clouds become larger, more dense, and have more distinct outlines. At this stage it is usually obvious that their smooth bases all lie at about the same level.

Although we will talk in more detail later about the way in which clouds form, it helps with the identification of cloud types if you know a little about what is happening and can relate it to what you see. In the case of cumulus clouds, they are the signs of

Cumulus fractus

bubbles of warm air (known as thermals), rising in the atmosphere. As they rise they cool, until they reach a level at which the water vapor within them condenses into tiny cloud droplets. As a result, the bases of cumulus cloud all occur at about the same level, and when there are a large number of individual clouds in the sky, it is readily apparent that the bases lie in a single plane. At any location, cloud base is often considerably higher in summer than in winter. This arises because the air is generally drier and warmer, so the thermals need to rise to a greater altitude for condensation to occur. In summer, too, the cloud base is frequently noticeably higher in the afternoon than in the early morning.

Early in the day, the thermals are small and rapidly mix with the surrounding air; they are also easily dispersed by the wind. Because of these processes, the cloud droplets soon evaporate. This is why early clouds are often ragged, small, and short-lived (cumulus fractus). As the day progresses, the clouds become larger and last longer. A characteristic feature of cumulus is that the individual heaps of cloud are generally fairly well separated from one another, even when they increase in size later in the day. Sometimes they form long lines of clouds (cloud streets) that stretch downwind, and there is a tendency for the clumps in an individual cloud street to merge. Even here, however, some gaps persist, and the variation in width is usually readily apparent.

Another feature of cumulus is that the tops of the clouds are rounded, and usually show signs that they are growing upwards. This is an indication that the air within them is still rising. When the movement ceases, the clouds start to decay, becoming ragged and eventually evaporating. When solar heating decreases in the late afternoon—unless other factors are at work—cumulus clouds start to decline in numbers, and become smaller. The exact behavior depends on atmospheric conditions and the amount of cloud that has built up during the day, but towards evening, only small ragged tufts of cloud may remain, similar to those seen early in the morning.

Although the color of cumulus clouds is usually described as white, like most clouds they may show a range of colors, depending on their background, and the amount of illumination (and also on the time of day). Generally the tops are white where they are illuminated by the Sun, but other parts appear grey or bluish

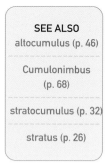

SEE ALSO

altocumulus (p. 46)

Cumulonimbus (p. 68)

stratocumulus (p. 32)

stratus (p. 26)

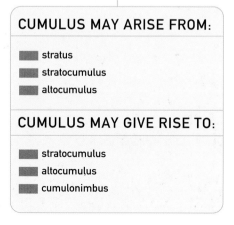

CUMULUS MAY ARISE FROM:

- stratus
- stratocumulus
- altocumulus

CUMULUS MAY GIVE RISE TO:

- stratocumulus
- altocumulus
- cumulonimbus

Cumulus mediocris

in tint, especially where the clouds are thin, usually where they are ragged and evaporating. When in shadow, they often appear black, especially when seen in silhouette against more distant, brightly illuminated clouds.

Apart from being created by thermals, cumulus may arise through the erosion of layers of stratus, stratocumulus or altocumulus. A layer of stratus over land frequently starts to lift in the morning, breaking up into stratocumulus, which may in turn finally give rise to cumulus. Cumulus are always a sign of a greater or lesser degree of instability in the atmosphere.

Cumulus frequently cease growing when they reach an inversion, where temperature increases with height. They are forced to spread out at this level, gradually blanketing the sky and becoming stratocumulus or, if higher, altocumulus. Under completely different conditions, they may become extremely vigorous, growing rapidly and clustering together to produce cumulonimbus clouds. These are shower clouds, with very specific, distinctive features.

Cumulus are water-droplet clouds, but the majority do not give rise to any rain. This is indicated by their flat bases: when any cloud produces precipitation, its base tends to become ragged. One species, cumulus congestus (to be described shortly), is a significant source of rain in the tropics and occasionally causes rainfall in temperate regions, particularly during the summer.

The four distinct species of cumulus are described on the following pages. As a class, cumulus are not generally confused with other clouds, although small cumulus in the distance may be difficult to distinguish from stratocumulus or altocumulus. Under such circumstances, binoculars often help to reveal more details and make it easier to estimate the clouds' height above the ground. If there are no signs that the bases are merging, and the tops show signs of upward growth, the clouds are considered to be cumulus. Cumulus congestus have many features in common with cumulonimbus, and discrimination between these two forms is slightly more difficult.

SEE ALSO

cloud formation (p. 86)

cloud streets (p. 92)

stability and instability (p. 188)

thermals (p. 188)

inversion (p. 187)

altocumulus (p. 46)

cumulonimbus (p. 68)

stratus (p. 26)

stratocumulus (p. 32)

Cumulus congestus

CUMULUS SPECIES

There are four species of cumulus, with very specific characteristics:

- CUMULUS FRACTUS (Cu fra) — BROKEN CUMULUS
- CUMULUS HUMILIS (Cu hum) — FLATTENED CUMULUS
- CUMULUS MEDIOCRIS (Cu med) — MEDIUM CUMULUS
- CUMULUS CONGESTUS (Cu con) — HEAPED CUMULUS

Cumulus fractus (Cu fra)

When clouds start to form on a clear day, initially there may be little more than slightly misty patches in the air. These give way to small clouds with ragged edges that change rapidly and continuously. These clouds are known as cumulus fractus. Some are little more than wisps of cloud, but because they are produced by rising thermals many have a roughly domed top. Similar clouds occur as other, larger cumulus decay, particularly at the end of the day.

Cumulus fractus: remnants of earlier cumulus clouds

This type of cloud may be difficult to distinguish from stratus fractus, but patches of the latter are normally an early stage in the development of a layer of true stratus, or are fragments left behind when that type of cloud has dispersed. Cumulus fractus is usually deeper, and appears both brighter and denser. Again, the tops are sometimes rounded, rather than relatively flat, as with stratus fractus.

Cumulus humilis (Cu hum)

Cumulus humilis are shallow clouds, with bases that are much wider than their vertical depth. The upper portions are gently rounded, but may sometimes display a distinctly flattened appearance. This species frequently occurs early in the day, having gradually grown from the initial cumulus fractus. If they continue to grow vertically, they will turn into cumulus mediocris. Often they arise, however, when conditions are tending to suppress convection, thus limiting upward growth, and giving rise to flatter tops. This happens in advance of the warm front of a depression, and also during anticyclonic conditions when air descends from higher levels, tending to prevent the thermals from rising beyond a certain height.

Cumulus humilis beneath high cirrus: an early indication of an approaching warm front

Cumulus mediocris (Cu med)

The next stage in the growth of cumulus clouds is cumulus mediocris. Here, there is obvious vertical growth, usually in the form of small, but clearly defined, rounded protuberances on the tops of the clouds. There may be a roughly triangular appearance, with one part of the cloud distinctly higher than other regions. The overall depth of the clouds remains less than or roughly equal to the width of the base. If there is a strong wind blowing, these clouds may begin to show signs of 'leaning' downwind, although this becomes much more marked in the next species: cumulus congestus.

Cumulus mediocris with some smaller cumulus humilis and more extensive cumuliform clouds in the background

Cumulus congestus (Cu con)

Cumulus that are growing vigorously and whose height is equal to or greater than the width of the base are known as cumulus congestus, often known as 'towering cumulus'. The upper regions usually appear dense and brilliantly white, and thus resemble a cauliflower. The tops are sharply outlined, and show no signs of a fibrous or striated structure, although they may fray out slightly as they decay. With strong winds, the towering clouds often lean downwind at a considerable angle.

Cumulus congestus may become sufficiently deep to produce rain, and when this starts to fall, the base of the cloud becomes ragged. Rainfall will be in the form of a short shower. If rain is falling from a cumulus congestus that is directly overhead, the cloud might be confused with altostratus or nimbostratus, but these types generally give rise to prolonged rain. A greater possibility of confusion arises with cumulonimbus

clouds, because these actually develop from cumulus congestus. Their distinctive characteristics are described later, but when a cloud is overhead they remain invisible. Conventionally, if there is neither hail nor lightning and the associated thunder, and if the characteristic features of a cumulonimbus such as an anvil or cirrus plume are invisible, the cloud is regarded as being a cumulus congestus.

Cumulus congestus: the cloud tower in the center is clearly still growing vigorously

CUMULUS VARIETY

SEE ALSO

anticyclone (p. 182)

depression (p. 180)

cloud streets (p. 92)

altocumulus (p. 46)

cumulonimbus (p. 68)

nimbostratus (p. 44)

stratus fractus (p. 30)

There is just one variety of cumulus that may be encountered: cumulus radiatus (Cu ra), when the individual clouds are arranged in distinct lines. These, known as cloud streets, will be discussed later, because they form under specific conditions.

Well-defined cloud streets (cumulus radiatus) beneath increasing high cirrus

LOW CLOUDS
STRATUS (St)

Stratus is another familiar cloud, although one that has few really distinctive features. It is a low, grey or blue-grey blanket of cloud, consisting of water droplets. It may hide the tops of high buildings and often descends to ground level, when it would be described as mist or fog. The base is often soft, rather than ragged, but there is little precipitation.

CHARACTERISTIC FEATURES

- Low, grey layer, with a fairly featureless base.
- One species (fractus) always appears ragged.
- Generally no precipitation, but may occasionally produce small quantities of drizzle, a few snowflakes or small ice crystals.

Stratus cloud is the lowest of all clouds (only rarely is its base higher than 1,700 ft.) and frequently associated with fog. In fact, there is no difference between stratus and fog, merely that the base of the former does not descend to ground level. Status is generally fairly uniform in its density, and may range from being thin enough for the outline of the Sun or Moon to be distinctly visible through it (in the variety known as stratus translucidus), to being so thick (stratus opacus) that there are no signs of either the Sun or Moon. When dense, it may take on a dark, blue-grey appearance and seem quite menacing. The base may appear soft (especially if one is close to it), but it is normally fairly well-defined and may show distinct undulations (stratus undulatus). As with other layer clouds, stratus may occur in patches, as well as a layer that covers the whole sky. Seen from above (from a hill-top or an aircraft, for example), it has a smooth, and relatively featureless surface.

Stratus cloud normally consists of water droplets, and if the Sun or Moon is visible, it may be surrounded by a corona. On extremely rare occasions, when temperatures are very low, stratus may contain ice crystals, which (again rarely) may give rise to a halo.

The most common method of formation occurs when moist air is carried over a cold surface. (The process of horizontal transport of air or other atmospheric properties is known as advection, described in more detail later.) Whether fog or stratus forms depends on a delicate balance between various factors, the most important of which are wind speed (with its resulting turbulence) and the temperature difference between the air and the surface. At low speeds, with a slow drift of air, there is lit-

tle turbulence, and even with a small temperature difference, the lowest layer of air cools first, leading to the formation of ground fog. At slightly higher speeds (around 6–12 mph), turbulence spreads the cooling through a greater depth of air. Stratus begins to form at the top of the mixed layer and, if mixing and cooling continues, may spread down towards the ground. The tops of hills may be in clear air above the cloud. If the wind speed is too high, the mixing (and cooling) occurs within a deep layer, and no cloud at all may form. If, however, there is a very large temperature difference between the air and the surface, stratus may occur even at high wind speeds, leading, for example, to the notorious 'Force 10 fog' over the Shetlands.

This stratus cloud, shrouding the hilltops, clearly shows that it is identical to thick fog.

Favorable conditions for the formation of stratus are found when a relatively warm, moist (maritime) air stream blows across any cold surface, such as a cold sea, or a thawing snow surface. The stratus (advection stratus) that forms over the sea may be carried onto nearby coasts and sometimes well

Stratus with slight undulations above a roadside weather station in Shetland

inland. The term 'haar' that was originally applied to low stratus of this type around Orkney is now used for both advection stratus and advection fog anywhere on the eastern coast of Scotland and northern England. It occurs most frequently in spring and early summer, when a relatively warm easterly air stream crosses the cold waters of the North Sea. The cloud generally first occurs as the broken stratus fractus species, and gradually thickens to become stratus nebulosus.

Under relatively still conditions, because of the way in which stratus forms, the tops of hills and mountains are often in bright sunshine above the layer of cloud. However, stratus also frequently occurs with stronger winds when moist air is forced to rise over hills or mountains, whose tops become shrouded in cloud, while the lower slopes, valleys and neighboring plains are cloud-free. Similar lifting of a layer of humid air behind the warm front of a depression can also lead to extensive stratus cover.

Stratus is frequently encountered when fog that has formed overnight begins to lift in the morning. When the fog is relatively thin, radiation from the Sun penetrates through it and heats the ground. This heat is transferred to the lowest layer of air, and some time after sunrise, the fog starts to disperse, its base lifting away from the surface, and transforming it into low stratus. A similar process occurs if the wind carries the cloud over a warmer surface. The cloud may also lift if the

STRATUS MAY ARISE FROM:

■ fog
■ stratocumulus

wind rises, causing greater turbulence, and mixing throughout a large volume of air. The cloud may disperse completely, but frequently fragments (stratus fractus) persist in valleys and along mountainsides.

Stratus may sometimes develop when stratocumulus loses its distinctive structure of individual cells or rolls. This happens when the shallow convection, responsible for the structure, ceases. The base of the cloud becomes indistinct, with no recognizable pattern of light and dark elements, and may lower towards the surface. Conversely, if convection begins within stratus, or if the wind carries it over high ground, forcing variations in its altitude, the cloud may break up into stratocumulus.

Distinguishing stratus from other cloud types

At a distance, stratus may be difficult to distinguish from stratocumulus, because the latter's structure is difficult to see. If any rounded heads or turrets are visible on the top of the layer it is

STRATUS MAY BE CONFUSED WITH:

- stratocumulus
- nimbostratus
- cirrostratus

safe to assume that it is stratocumulus. In addition, the edges of a stratus layer usually taper away gradually, unlike the more distinct, roughly vertical edges seen on stratocumulus. Seen from above, stratus does not display the more-or-less regular pattern of breaks found in stratocumulus.

Stratus is most likely to be mistaken for nimbostratus, but generally has a more uniform base. Nimbostratus is not only frequently more ragged and uneven, but it also produces persistent and fairly continuous precipitation. By contrast, stratus is essentially 'dry', and merely gives a slight drizzle. Nimbostratus is also generally preceded by higher clouds, particularly altostratus, and always obscures the Sun, whereas the outline of the Sun and Moon may be seen through thin stratus. Unlike the higher altostratus, stratus does not blur the outline of the Sun or Moon with a 'ground-glass' effect.

Although stratus sometimes shows undulations (in the variety known as stratus undulatus), the lack of individual elements in the form of clumps, pancakes or rolls distinguishes it from stratocumulus. It is fairly unlikely to be mistaken for the higher cirrostratus (p. 60), unless in a very thin layer, but it would then probably show a corona around the Sun or Moon, rather than the halo effects seen in cirriform clouds. On occasions, stratus may assume a coarse fibrous appearance, somewhat like cirrus but the fibers are denser, and are very dark (as well as being obviously relatively close to the ground).

SEE ALSO

altostratus (p. 40)

cirrostratus (p. 60)

cirrus (p. 54)

corona (p. 104)

convection (p. 86)

fog (p. 142)

halo (p. 116)

nimbostratus (p. 44)

stratocumulus (p. 32)

warm front (p. 179)

STRATUS SPECIES

There are two species of stratus:

STRATUS FRACTUS (St fra) – BROKEN STRATUS

STRATUS NEBULOSUS (St neb) – FEATURELESS STRATUS

When stratus begins to form, it usually first occurs as individual, ragged scraps of cloud. The separate fragments of stratus fractus gradually thicken and merge into a layer of stratus nebulosus. The latter is often remark-

Featureless stratus nebulosus merely shows an indistinct image of the Sun

Stratus may break up into stratus fractus

ably uniform and featureless, showing little variation in density and shading. Stratus fractus reappears when a layer of stratus starts to disperse.

Stratus fractus might be confused with cumulus fractus, but the different circumstances under which the two species occur is normally sufficient for them to be distinguished easily. One particular form of stratus, known as pannus, occurs beneath clouds that are producing rain or other precipitation. This type of accessory cloud is described later.

SEE ALSO

corona (p. 104)

pannus (p. 72)

STRATUS VARIETIES

There are three varieties of stratus, which are all easily recognized:

- STRATUS OPACUS (St op) – OPAQUE STRATUS
- STRATUS TRANSLUCIDUS (St tr) – TRANSLUCENT STRATUS
- STRATUS UNDULATUS (St un) – UNDULATING STRATUS

As the names suggest, stratus opacus is thick enough to hide the Sun or Moon, whereas stratus translucidus is thinner, permitting the bodies' outlines to be seen. Stratus translucidus may thus give rise to a corona around the Sun or Moon. The base of stratus undulatus shows parallel waves which are commonly 1,600–3,000 ft. apart. Generally such waves are approximately at right-angles to the direction of the wind.

Stratus undulatus with an uncommon fibrous character

Stratocumulus is another low, layer cloud, but one that shows far more variation and interest than stratus. It is an extremely common form, particularly over the oceans.

CHARACTERISTIC FEATURES

- Low cloud, white to dark grey, broken into individual heaps and masses.
- Occurs as an extensive layer or smaller patches.
- Sometimes produces light precipitation.

Stratocumulus is a very varied cloud, which shows considerable differences in tint from white to grey or dark blue-grey. It always consists of individual cloud masses, which are in the form of rounded masses, rolls, or pancakes, with very pronounced shading. Some of these may be merged into larger bodies of cloud that are, in effect, patches of stratus. Overall, the individual cloud elements are fairly regularly arranged, and most appear more than 5° across, measured 30° above the horizon. Stratocumulus may sometimes produce light precipitation in the form of rain or snow pellets, and occasionally some snow crystals or snowflakes.

The base of a stratocumulus layer is usually fairly well defined—which may be expected from a cloud that produces little precipitation at the ground—but the upper surface (if visible) is frequently very uneven and ragged. The layer is generally fairly shallow, but is often extremely extensive, easily covering the whole of the sky that is visible from any one point. It is quite common for stratocumulus to occur as more than one layer at different heights (in the variety known as stratocumulus duplicatus).

There are two main methods of formation and, interestingly, the mechanisms are essentially the opposite of one another. One method occurs when a humid, stable layer of air is forced to rise, because it is lifted within a depression system, when it encounters high ground, or when heating or an increase of wind speed causes a layer of fog or stratus to rise. The layer of cloud that is thus initially produced, breaks up when shallow convection begins within the layer. This often starts when the top of the cloud radiates heat away to space (even during the daytime). Air subsides in the cooler areas, initiating the convection, and breaking up the cloud layer.

The second method of formation occurs when there is a relatively weak convection rising from the ground, and cumulus clouds form in a shallow layer, or are forced to spread out sideways by an inversion. If the convection is stronger in some areas than others, some towers may break through the inversion and the stratocumulus layer. Sometimes, when convection becomes particularly strong as the day progresses, stratocumulus may even develop into larger cumuliform clouds, such as cumulus congestus and cumulonimbus, although these may be difficult to recognize from the ground, being hidden by the general layer of cloud, and perhaps only occasionally glimpsed through larger openings. Seen from an aircraft above the layer, they are, of course, readily visible.

Areas of stratocumulus are often found both ahead of, and behind major shower or thunderstorm systems. They form ahead of the system where air is being drawn into it, and gradually decay behind it.

STRATOCUMULUS MAY ARISE FROM:

- break-up of stratus cloud
- spreading out of cumulus clouds
- remnants of storm systems

STRATOCUMULUS MAY DEVELOP INTO:

- scattered cumulus
- a continuous layer of stratus
- altocumulus

Heavy stratocumulus stratiformis over the River Seine in Paris

A stratocumulus layer generally decays with increasing wind speed and turbulence, or by the development of strong convection. Both result in the layer becoming more and more fragmented, until it breaks up into isolated cumulus clouds. Conversely, if the wind drops, and convection weakens, such as towards dusk, stratocumulus may thicken to become more or less an unbroken layer of stratus.

Stratocumulus, like altocumulus, displays a wide range of varieties, with the individual elements varying considerably in their shape and overall form. Sometimes the cloud-masses may be distinct, relatively restricted clumps that obviously have a fairly considerable depth, and at others they may spread out into broad, thin 'pancakes', with clear sky between them (stratocumulus perlucidus). On rare occasions, the appearance may be reversed, giving a network of cloud, surrounding large, clear holes in the layer (stratocumulus lacunosus).

More frequently, stratocumulus forms billows, distinct rolls of cloud that lie approximately at right-angles to the wind direction. The cloud may be fairly thick, with only narrow breaks, so that little more than the undulating base is seen, or there may be distinct breaks with clear sky between the billows. A layer that consists of this variety (stratocumulus undulatus) often has an extremely striking appearance. The formation of

STRATOCUMULUS MAY BE CONFUSED WITH:

▩ altocumulus

▩ stratus

▩ nimbostratus

such billows, which also occur with other cloud types, is discussed in more detail later.

Stratocumulus also frequently merges into elongated masses of cloud, aligned roughly parallel with the wind direction. Sometimes these assume the appearance of long, well-defined rolls or lens-shaped patches of cloud, relatively isolated from one another, that are visible outlined beneath a higher layer of cloud, such as altostratus or altocumulus.

Although stratocumulus produces little precipitation at the surface, it may have distinct virga (fallstreaks) beneath it, where rain is falling from the cloud and evaporating in the intervening layer. There may sometimes be distinct pouches (mamma) on the underside of the cloud layer. Although funnel clouds (tuba) have been reported beneath stratocumulus, these probably originated from embedded cumulus congestus or cumulonimbus clouds that were invisible from the ground.

Distinguishing stratocumulus from other cloud types

Thin stratocumulus may produce a corona or iridescence, but these are less common than in altocumulus. Under extremely cold conditions—not often encountered at temperate latitudes—the virga beneath stratocumulus may consist of ice crystals, which may then give rise to a halo or halo phenomena such as a parhelion (mock sun). Such phenomena are more often associated with cirrostratus or other cirriform clouds, but stratocumulus is always much thicker, and shows distinct masses or rolls of cloud, so there is little likelihood of making a mistake.

Stratocumulus is most likely to be confused with altocumulus that shows dark shading. If, however, most of the cloud elements, 30º or more above the horizon, have a width of 5º or more, then the cloud is stratocumulus. A stratus layer shows little structure, although it may, as the day progresses and convection increases, break up into stratocumulus. Nimbostratus does not exhibit any regular structure, is generally far more ragged in appearance and—the deciding factor—is accompanied by considerable precipitation, unlike stratocumulus which produces an insignificant amount.

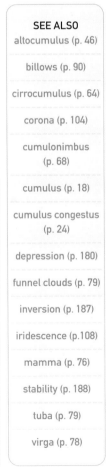

SEE ALSO

altocumulus (p. 46)

billows (p. 90)

cirrocumulus (p. 64)

corona (p. 104)

cumulonimbus (p. 68)

cumulus (p. 18)

cumulus congestus (p. 24)

depression (p. 180)

funnel clouds (p. 79)

inversion (p. 187)

iridescence (p.108)

mamma (p. 76)

stability (p. 188)

tuba (p. 79)

virga (p. 78)

STRATOCUMULUS SPECIES

There are three species of stratocumulus:

▢ STRATOCUMULUS CASTELLANUS (Sc cas) — TURRETED STRATOCUMULUS

▢ STRATOCUMULUS LENTICULARIS (Sc len) — LENTICULAR STRATOCUMULUS

▢ STRATOCUMULUS STRATIFORMIS (Sc str) — EXTENSIVE STRATOCUMULUS

In stratocumulus castellanus, distinct heads rise from an elongated line of cloud with a relatively smooth base. These cloud turrets are a sign of instability at that level, and they may progress to form cumulus congestus (towering cumulus) or even cumulonimbus clouds. They are not, however, as certain an indicator of forthcoming unsettled weather as the similar, but higher, altocumulus castellanus clouds.

Stratocumulus lenticularis clouds are smooth, lens- or almond-shaped clouds that develop as a result of undulations in the airflow caused by hills or mountains. Their name is somewhat misleading, because cumulus clouds are normally associated with convection and instability, but stratocumulus lenticularis (and the similar altocumulus lenticularis and cirrocumulus lenticularis) occur when a stable layer of air is forced into a wavelike motion. Such wave clouds are discussed in greater detail later, but it should be noted here that—unlike most clouds—they remain stationary in the sky all the time that the wind speed and direction remain constant.

> High stratocumulus castellanus indicating instability at cloud level, which might promote the formation of later showers

When the cloud occurs as an extensive layer, rather than just isolated, individual masses, it is described as being of the species stratocumulus stratiformis. This form of stratocumulus (and the corresponding altocumulus and cirrocumulus stratiformis) often exhibit great variations in size, shape, color, and shading of the individual cloud elements, giving an extremely interesting aspect to the sky. Such extensive sheets of cloud

are typical when the air is subject to slow uplift, particularly near the fronts in depression systems. The exact nature of the cloud layer may be indicated by specifying the particular cloud variety that is present: lacunosus, perlucidus, undulatus, etc.

STRATOCUMULUS VARIETIES

A total of seven varieties of stratocumulus are officially recognized:

- STRATOCUMULUS DUPLICATUS (Sc du) — MULTIPLE STRATOCUMULUS LAYERS
- STRATOCUMULUS LACUNOSUS (Sc la) — STRATOCUMULUS WITH LARGE CLEAR HOLES
- STRATOCUMULUS OPACUS (Sc op) — OPAQUE STRATOCUMULUS
- STRATOCUMULUS PERLUCIDUS (Sc pe) — STRATOCUMULUS WITH CLEAR GAPS BETWEEN THE ELEMENTS
- STRATOCUMULUS RADIATUS (Sc ra) — STRATOCUMULUS IN PARALLEL LINES
- STRATOCUMULUS TRANSLUCIDUS (Sc tr) — TRANSLUCENT STRATOCUMULUS
- STRATOCUMULUS UNDULATUS (Sc un) — STRATOCUMULUS WITH UNDULATIONS

Stratocumulus is classed as stratocumulus duplicatus when it occurs as patches or sheets at different levels. Such individual layers frequently merge into a single cloud mass in one particular part of the sky.

As previously described for stratus clouds, the opacus variety implies that the cloud is too dense and opaque for the position of the Sun or

Moon to be determined. With stratocumulus translucidus, the cloud is thin, and the outline of the Sun or Moon may be seen through the cloud, occasionally accompanied by a corona.

Stratocumulus perlucidus and stratocumulus lacunosus might be described as 'mirror images' of one another. The perlucidus variety is by far the most common, and describes the appearance when the cloud is broken up into individual flat 'pancakes' like crazy paving. This arises when the top of a sheet of cloud radiates heat away to space. A particular form of shallow convection begins within the cloud layer, causing the cloud in the center of each convection cell to thicken, and that around the outside to thin, or disperse entirely. Both the higher altocumulus and cirrocumulus exhibit the same sort of structure, but there is an approximate relationship between the diameter of the cells and height. The higher the cloud, the smaller the cells. Not only is stratocumulus lower, so the

Stratocumulus duplicatus (BELOW)

Stratocumulus perlucidus (BOTTOM)

cells appear larger, but each cell covers a larger area than the higher cloud types.

Stratocumulus lacunosus also begins as an essentially even layer of cloud. Here, however, the convection mixes drier air from a higher or lower level into the moist layer. The result is that the cloud in the centers of the convection cells is eroded, leaving a network or honeycomb of cloud surrounding cloud-free holes. This form is much rarer than the perlucidus variety.

As mentioned earlier, the individual billows in stratocumulus undulatus may be thick and close together, so that only the undulating base is visible, or they may be well separated with clear blue sky between them.

The final variety, stratocumulus radiatus, occurs when the cloud is arranged in long rolls. The clouds therefore appear to diverge from a single point in the sky in the upwind direction and, if the sheet of cloud is large enough, to converge on the opposite side of the sky. The varieties perlucidus, undulatus, and radiatus frequently occur together, to give a beautifully patterned sky, commonly called a 'mackerel sky', although this term should really be reserved for cirrocumulus.

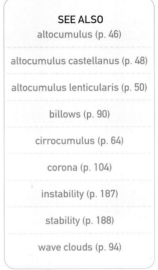

SEE ALSO

altocumulus (p. 46)

altocumulus castellanus (p. 48)

altocumulus lenticularis (p. 50)

billows (p. 90)

cirrocumulus (p. 64)

corona (p. 104)

instability (p. 187)

stability (p. 188)

wave clouds (p. 94)

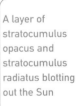

A layer of stratocumulus opacus and stratocumulus radiatus blotting out the Sun

MIDDLE CLOUDS
ALTOSTRATUS (As)

Moving to the middle level, we have altostratus and nimbostratus (two stratiform cloud types) and altocumulus (a cumuliform type).

CHARACTERISTIC FEATURES

- A layer cloud, grey or bluish grey, generally extremely extensive, uniform or striated in appearance.
- Usually gives rise to considerable precipitation, which may reach the ground.

Altostratus has sometimes been described as a boring cloud. It displays little variation and, like nimbostratus, it is not subdivided into different species, unlike the other eight types. It is always light to dark grey or bluish grey in tint, and is usually extremely extensive and thick, often some hundreds of miles across and thousands of feet thick. Naturally, smaller or thinner patches also occur. Another characteristic is that it tends to be remarkably uniform (and featureless) in appearance, although it occasionally shows a pronounced striated or fibrous structure.

Altostratus is a mixed cloud, consisting of both tiny cloud droplets and minute ice crystals, together with larger raindrops and snowflakes. For this reason it does not display any halo phenomena, but when the cloud is thin, it may allow the Sun (and very occasionally the Moon) to be seen through it, with a highly characteristic appearance, as if through ground glass. When thick, however, it completely masks the Sun, so that not even its position may be determined. The light is diffuse, and objects on the ground do not cast any shadows. The edges of a sheet of altostratus, particularly when

Altostratus ahead of an advancing warm front, above stratus fractus

decaying, may consist of just water droplets. These regions sometimes display slight iridescence or portions of a corona around the Sun or Moon.

Altostratus normally produces a lot of precipitation, which nearly always lasts for a long time, and which may reach the ground as rain, snow, or ice pellets. Virga (fallstreaks) are common, and are often responsible for the striated appearance that altostratus exhibits. When precipitation from altostratus evaporates in lower layers, they are cooled and become humid. Turbulence within them may lead to the development of the accessory cloud known as pannus (stratus fractus), which may gradually increase and grow into a more-or-less continuous layer.

ALTOSTRATUS DEVELOPS FROM:

 uplift of a moist, stable layer

cirrostratus

nimbostratus

the spreading out of cumulonimbus clouds

altocumulus clouds

ALTOSTRATUS MAY DEVELOP INTO:

altocumulus

nimbostratus

As with the other stratiform clouds, altostratus is most frequently produced by the gentle uplift of a layer of moist air. This commonly occurs ahead of the advancing warm front of a depression. In many cases, it is preceded by a layer of cirrostratus that gradually thickens and lowers and eventually turns into altostratus. Behind a frontal system, altostratus often arises through the thinning of the thicker, lower, rain-bearing nimbostratus cloud.

At low latitudes, altostratus may also be produced when the middle or upper portions of cumulonimbus clouds or large thunderstorm systems spread out at an appropriate level. Fragments of altostratus may sometimes occur ahead of such systems, but more commonly trail behind them. (At higher latitudes, some altostratus may be produced, but stratus or stratocumulus tends to be more common under similar circumstances.)

When altocumulus clouds produce extensive virga, the cloud cover may increase sufficiently for the cumuliform nature to be lost, giving way to a sheet of altostratus.

Distinguishing altostratus from other cloud types

Dense cirrus (cirrus spissatus) may sometimes resemble altostratus, but the latter normally has a far greater extent, and tends to be darker in appearance. Cirrostratus often thickens and lowers eventually becoming thin altostratus. If objects on the ground cast shadows, or if any halo phenomena are present, then the cloud is cirrostratus.

Like any layer cloud, altostratus may have breaks, which might cause it to be confused with altocumulus or the lower stratocumulus. Generally, altostratus has a far more uniform appearance than either of these types, and certainly does not exhibit any regular pattern of small cloud masses.

Low altostratus somewhat resembles nimbostratus, but the latter is generally a much denser cloud, without breaks or thinner patches that might permit the Sun or Moon to be glimpsed through it. Altostratus is also lighter in color, and tends to have a more ragged base. If there is any doubt, and no precipitation is reaching the ground, then such a middle-level cloud is classed as altostratus. It may be distinguished from the much lower stratus by the fact that it is a much darker grey, and never appears white like thin stratus. More importantly, stratus never exhibits the 'ground-glass' effect found in altostratus.

As with stratus and cirrostratus, a layer of altostratus may break up into individual elements, thus becoming altocumulus, if shallow convection begins within the layer. Individual altocumulus elements tend to be smaller in diameter than those of stratocumulus, and larger than those encountered in the higher cirrocumulus. The base of an altostratus layer sometimes develops downward bulges (mamma) which are a sign of descending, cold air currents within the cloud.

Altostratus frequently lowers and thickens to become nimbostratus. The sequence: cirrostratus, altostratus, nimbostratus is highly characteristic of the approaching warm front of a depression.

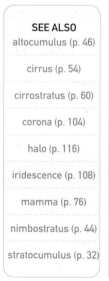

ALTOSTRATUS MAY BE CONFUSED WITH:

- cirrus
- cirrostratus
- altocumulus
- stratocumulus
- nimbostratus
- stratus

SEE ALSO

altocumulus (p. 46)

cirrus (p. 54)

cirrostratus (p. 60)

corona (p. 104)

halo (p. 116)

iridescence (p. 108)

mamma (p. 76)

nimbostratus (p. 44)

stratocumulus (p. 32)

ALTOSTRATUS VARIETIES

As mentioned, there are no altostratus species, but there are five varieties:

- ALTOSTRATUS DUPLICATUS (As du) — MULTIPLE ALTOSTRATUS LAYERS
- ALTOSTRATUS OPACUS (As op) — OPAQUE ALTOSTRATUS
- ALTOSTRATUS RADIATUS (As ra) — ALTOSTRATUS IN PARALLEL LINES
- ALTOSTRATUS TRANSLUCIDUS (As tr) — TRANSLUCENT ALTOSTRATUS
- ALTOSTRATUS UNDULATUS (As un) — UNDULATING ALTOSTRATUS

SEE ALSO
cirrocumulus (p. 64)

stratocumulus
(p. 32)

Altostratus opacus is so thick that it completely hides the Sun, and not even its position is detectable. Altostratus translucidus, on the other hand, is sufficiently transparent for the disk of the Sun (and sometimes of the Moon as well) to be seen through it, although the outline is obscured as if seen through ground glass.

Altostratus duplicatus indicates altostratus that occurs as sheets or patches at more than one height, which, as with stratocumulus, may merge in places into a single layer.

Occasionally altostratus forms in long rows, which may be parallel to, or at right-angles to the wind direction. In the first case, because the cloud appears to radiate from a single point in the sky, this variety is known as altostratus radiatus. (If the cloud sheet is very large, the rows may appear to converge to what are known as 'radiation points' on opposite sides of the sky.) In the second variety, altostratus undulatus, the undulations lie at right-angles to the wind direction. Both varieties may assume a very striking appearance when illuminated by low-angle illumination at sunrise or sunset.

Altostratus translucidus (TOP)
Altostratus radiatus (BOTTOM)

NIMBOSTRATUS (Ns)

Nimbostratus clouds are all too familiar. They are the clouds that produce the heavy and seemingly endless rain or snow that falls from depressions. Although classed as a middle-level cloud, its base is generally low, sometimes reaching down to within a short distance of the surface.

CHARACTERISTIC FEATURES

- A thick, dark cloud that completely masks the Sun.
- Forms in extremely extensive sheets with a low base, and gives rise to more-or-less continuous precipitation that persists for a long time.

Of the ten major cloud types, nimbostratus displays the least variation. In common with altostratus, it is not subdivided into species, but unlike that cloud type, neither does it have any recognized varieties. It is just a grey, often dark grey, cloud layer that produces more-or-less continuous rain or snow. It is always thick enough to completely hide the Sun (and the Moon). Because of its density, it never gives rise to any optical phenomena such as haloes or coronae. Its base appears soft or diffuse because of the precipitation from the cloud layer. This base is often obscured by the ragged pannus (cumulus fractus or stratus fractus) that forms in the saturated air beneath the main body of nimbostratus. Although individual small areas of cloud may not be producing precipitation at any one moment, the rain or snow is generally widespread. It is usually organized into bands, giving long periods of precipitation broken by spells when there may be a temporary respite.

The composition of nimbostratus may vary greatly depending on temperature. It sometimes consists entirely of water droplets of various sizes (i.e., both cloud droplets and raindrops) and at others of ice crystals and larger snowflakes. On occasions it may be mixed, with water droplets and ice crystals present at the same level.

The most widespread layers of nimbostratus arise through the slow uplift of moist air in a depression, typically at the advancing warm front. Here, it is very frequently preceded by altostratus, which thickens and lowers towards the surface, almost imperceptibly turning into nimbostratus. On slow-moving

A layer of nimbostratus beginning to break up with the passage of a cold front

NIMBOSTRATUS MAY BE CONFUSED WITH:

- altostratus
- altocumulus and stratocumulus
- cumulonimbus
- stratus

occluded fronts, nimbostratus sometimes produces days of persistent rain or snow and often causes widespread flooding or record-breaking snowfalls.

More rarely, nimbostratus results from the thickening of altocumulus or stratocumulus. Large cumulonimbus clouds or thunderstorm systems may sometimes spread out and give rise to nimbostratus, but this covers a smaller area than the vast sheets found in depressions. On very rare occasions, nimbostratus may also arise from large, rain-bearing cumulus congestus clouds. Again, the extent of any resulting layer is greatly restricted.

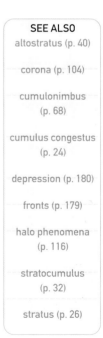

Nimbostratus with pannus bringing persistent rain to the mountains of the Peak District

Distinguishing nimbostratus from other cloud types

Nimbostratus is distinguished from altostratus by its thickness. It is never thin enough to allow the position of the Sun or Moon to be determined, whereas only the thicker portions of altostratus hide those bodies completely. By convention, if it is impossible to determine whether the cloud is nimbostratus or altostratus—such as during a dark night—the cloud is described as nimbostratus if precipitation is reaching the ground.

Thick altocumulus and stratocumulus may be distinguished from nimbostratus by the fact that they display a more-or-less regular pattern of individual cloud elements, and also generally have a clearly defined base.

When a cumulonimbus is overhead, and its distinctive features cannot be seen there is a chance that it might be mistaken for nimbostratus. Generally, however, it will be clear from the prevailing conditions, which type is the most probable, although there may be some doubt when close to the cold or occluded fronts in a depression system. If the cloud is accompanied by lightning, thunder or hail, then it is defined as being cumulonimbus.

Thick stratus could be confused with nimbostratus, but the latter gives rise to generally heavy precipitation, whereas stratus produces light drizzle or, under cold conditions, a tiny amount of precipitation in the form of snow grains or tiny ice crystals. In general, the base of nimbostratus is also more ragged and uneven than that of stratus.

SEE ALSO

altostratus (p. 40)

corona (p. 104)

cumulonimbus (p. 68)

cumulus congestus (p. 24)

depression (p. 180)

fronts (p. 179)

halo phenomena (p. 116)

stratocumulus (p. 32)

stratus (p. 26)

Altocumulus clouds produce some of the most striking and varied skies. This is not only because there are four distinct species and no fewer than seven varieties, but also because, being a broken cloud that occurs at medium altitudes, it may undergo dramatic changes in lighting at sunrise and sunset.

CHARACTERISTIC FEATURES

- A white to pale grey cloud, occurring (generally in a distinct layer) as heaps, rolls or pancakes, with darker shading.
- Only on rare occasions does precipitation reach the ground.

Altocumulus has much in common with the lower stratocumulus and higher cirrocumulus. It occurs in the same type of rounded masses of cloud, with clear sky between them. It is distinguished from those two types by the size of the individual elements. A cloud is classed as altocumulus if, 30° or more above the horizon, the average size of one of the cloud masses is less than 5° across, but larger than one degree. Another feature is the strength of the shading, which is distinct, but not extremely dark. Again, it comes between the stratocumulus (which has dark shading) and cirrocumulus (which shows no shading at all).

Generally, altocumulus occurs as an extensive sheet of cloud, consisting of more-or-less regular smaller cloud elements. As with stratocumulus and cirrocumulus, these may be in the form of small rounded clumps; larger, flatter 'pancakes', or extensive rolls of cloud. Blue sky is normally clearly visible between the individual elements, which always show some

The well-defined gaps between the cloud elements show that this layer should be classified as altocumulus stratiformis perlucidus

shading. Various other forms such as smooth, lens-shaped clouds; tufts with ragged bases (often with distinct virga or fallstreaks); and cloud turrets that rise from a common base are also encountered and are described in more detail under the various species. Because of its height, many more of the individual cloud elements of altocumulus are visible than with the lower stratocumulus, so the sky often appears full of innumerable masses of cloud.

Altocumulus varies considerably in its transparency: it may completely mask the Sun and Moon, or be thin enough for their positions to be determined quite easily. It often consists solely of fine water droplets (frequently in the supercooled state) but, at lower temperatures, it may also be a mixed cloud that contains ice crystals. As a result, altocumulus clouds may display optical phenomena associated either with water droplets (such as coronae or iridescence) or with ice crystals (such as parhelia and sun pillars). When the individual elements are close together, the base of thick altocumulus often displays a lumpy appearance, which may become extremely marked when illuminated at a shallow angle by the rays of the rising or setting Sun.

As with stratocumulus and cirrocumulus, altocumulus forms either through the break-up of existing cloud, particularly altostratus, or through the uplift of a humid layer of air. Once again, if a sheet of unbroken cloud loses heat from its upper surface to space, it tends to break up into individual convection cells. The cloud thickens in the center of each cell, and dissipates around the edges, giving rise to altocumulus. In addition, broken altocumulus is frequently found around the fringes of a more extensive sheet of altostratus. For physical reasons, the diameters of the individual cells are smaller than in stratocumulus, and the cloud thickness is generally less.

In a somewhat similar fashion, a decaying sheet of nimbostratus may leave altocumulus clouds behind it. This frequently occurs with the passages of fronts in a depression system. Sometimes deep stratocumulus may break up with the higher levels turning into altocumulus.

Scattered individual clumps of cloud may occur with the general uplift of a humid layer of air. On occasions, altocumulus may be formed through the thickening of individual patches of cirrostratus. The most distinctive form is created when the moist layer is forced into regular wave motion by underlying hills or mountains. The resulting altocumulus

ALTOCUMULUS MAY FORM:

- from altostratus
- from nimbostratus
- from stratocumulus
- by uplift of a layer of humid air
- by the spreading out of cumulus or cumulonimbus

lenticularis will be described later. These, together with billows, which arise when there is wind shear at the appropriate level, and are aligned approximately at right-angles to the wind direction, produce some of the most dramatic cloud effects.

Just as cumulus clouds may spread out to produce a layer of stratocumulus, so they may also reach a stable layer at a higher level and expand to create a sheet of altocumulus, thus producing altocumulus cumulogenitus (Ac cugen). Altocumulus also frequently forms ahead of, and trails behind, major thunderstorm (cumulonimbus) systems.

Distinguishing altocumulus from other cloud types

Small altocumulus with long virga may show a considerable resemblance to cirrus clouds. If there is a distinct clump of cloud (known as the generating head) without a fibrous appearance, then the cloud may be classed as altocumulus.

Confusion is probably most likely with cirrocumulus. Note that this type does not show any shading at all, so any cloud with shading is automatically classed as altocumulus. On the rare occasions when altocumulus has little shading, provided the individual cloud elements are more than 1 degree across (30° or more above the horizon), then cirrocumulus is ruled out. Similarly, although cirrocumulus can, on very rare occasions, display a corona, or iridescence, these phenomena are far more common in altocumulus.

It may sometimes be difficult to distinguish between thick altocumulus and altostratus, particularly the altostratus undulatus variety. Generally, if there is evidence of a regular pattern of masses or rolls, rather than just undulations of the lower surface, the cloud is altocumulus.

Similarly, dark altocumulus may be confused with stratocumulus. Here the distinction is made on the size of the individual elements. If the average size (again 30° or more above the horizon) is between 1 and 5°, the cloud is defined as altocumulus.

Some scattered clumps of altocumulus may resemble small cumulus. Generally, however, such small tufts of altocumulus show distinct trails (virga) and are usually much smaller in apparent size than the lower cumulus.

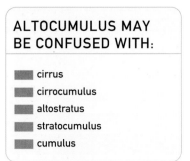

ALTOCUMULUS MAY BE CONFUSED WITH:

- cirrus
- cirrocumulus
- altostratus
- stratocumulus
- cumulus

ALTOCUMULUS SPECIES

There are four, very distinctive species of altocumulus:

- ALTOCUMULUS CASTELLANUS (Ac cas) – TURRETED ALTOCUMULUS
- ALTOCUMULUS FLOCCUS (Ac flo) – TUFTED ALTOCUMULUS
- ALTOCUMULUS LENTICULARIS (Ac len) – LENTICULAR ALTOCUMULUS
- ALTOCUMULUS STRATIFORMIS (Ac str) – SHEETS OF ALTOCUMULUS

Altocumulus castellanus (Ac cas)

Altocumulus castellanus shows distinct cloud towers or turrets rising from a fairly uniform line or layer of cloud. They are a sign that there is instability at that level. They are one of the indicators that vigorous convection may occur within the near future, with the chance of strong thunderstorm activity.

These altocumulus castellanus clouds occurred several hours before a major outbreak of thunderstorms

Altocumulus floccus (Ac flo)

Altocumulus floccus consists of small individual, more-or-less rounded tufts of cloud with ragged bases or trailing virga. They often appear in considerable numbers, giving an extremely striking appearance to the sky. Although normally quite widely separated in the sky, they sometimes occur in more closely packed lines. They are another sure sign of instability at that altitude and again are often an indication of thundery activity to come.

Altocumulus lenticularis (Ac len)

Clouds of the altocumulus lenticularis species are smooth lens- or almond-shaped clouds that hang stationary in the sky, often for extended periods. They arise

Well-defined and characteristic altocumulus floccus (top). Altocumulus lenticularis and lower cumulus clouds (bottom)

in the crests of waves that are produced when a stable layer of air is forced to rise over hills or mountain peaks. Their smooth outlines are particularly striking, and if they are examined in detail—with binoculars, for example—slight fluctuations show where they are forming on the leading edge and dissipating on the downwind (trailing) side. The accessory clouds known as pileus form by a somewhat similar mechanism.

Sometimes multiple humid layers may produce lenticular clouds one above the other, and this striking appearance is known as a 'pile d'assiettes' (French for 'pile of plates'). Under certain conditions, individual clouds may merge to form greatly elongated, smooth lines of cloud. Altocumulus lenticularis, like other wave clouds, tend to persist while the wind speed and direction remain constant.

Altocumulus stratiformis (Ac str)

Altocumulus that forms as an extensive layer, usually covering the whole of the sky, rather than just small patches, is known as altocumulus stratiformis. The fact that essentially all of the visible sky is covered by a regular pattern of cloudlets in a conspicuous layer often leads to a very striking appearance, but this may not persist for long as the lighting conditions change.

SEE ALSO

instability (p. 187)

pileus (p. 72)

virga (p. 78)

orographic clouds (p. 87)

wave clouds (p. 94)

An extensive layer of altocumulus stratiformis at sunset

ALTOCUMULUS VARIETIES

In parallel with the varieties of stratocumulus, there are seven distinct varieties of altocumulus:

- ALTOCUMULUS DUPLICATUS (Ac du) — MULTIPLE ALTOCUMULUS LAYERS
- ALTOCUMULUS LACUNOSUS (Ac la) — ALTOCUMULUS WITH LARGE CLEAR HOLES
- ALTOCUMULUS OPACUS (Ac op) — OPAQUE ALTOCUMULUS
- ALTOCUMULUS PERLUCIDUS (Ac pe) — ALTOCUMULUS WITH CLEAR GAPS BETWEEN THE ELEMENTS
- ALTOCUMULUS RADIATUS (Ac ra) — ALTOCUMULUS IN PARALLEL LINES
- ALTOCUMULUS TRANSLUCIDUS (Ac tr) — TRANSLUCENT ALTOCUMULUS
- ALTOCUMULUS UNDULATUS (Ac un) — ALTOCUMULUS WITH UNDULATIONS

Altocumulus quite commonly occurs at more than one distinct level, when it is known as altocumulus duplicatus. The clouds at these

different levels frequently merge into a single cloud mass over one or more parts of the sky.

As with most of the other cloud types, the opacus variety, altocumulus opacus, is so dense that no glimpses of either the Sun or Moon may be seen through it. Altocumulus translucidus, on the other hand, is thin enough for the position of these bodies to be clearly seen.

Altocumulus perlucidus is the most common variety, where the individual cloud masses are surrounded by narrow clear lanes. As with stratocumulus perlucidus and cirrocumulus perlucidus, this often arises when shallow convection begins in a previously unbroken layer of cloud—in this case, altostratus. Altocumulus lacunosus, by contrast, is much rarer, and consists of a network of cloud, surrounding large clear holes.

Altocumulus radiatus clouds that were clearly streaming from lower left

SEE ALSO
cloud formation
(p. 86)

wind shear (p. 188)

The final two varieties may also be considered as a related pair. In altocumulus radiatus, the individual cloudlets are arranged in lines, giving the impression, when viewed in the appropriate direction, of radiating from a point upwind, and converging on the opposite, downwind side. With altocumulus undulatus the rolls of cloud are arranged approximately at right-angles to the wind direction. Such a distribution is commonly the result of wind shear at cloud level. Both varieties often occur together—when there is an extensive sheet of altocumulus undulatus, it is almost certain to show the radiatus characteristics—and in combination with the perlucidus variety. This produces a beautiful, and often very striking, pattern of cloud elements. This combination is commonly called a 'mackerel sky', although the term should really be reserved for cirrocumulus.

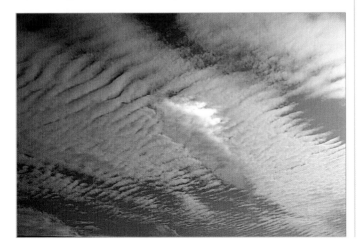

Altocumulus undulatus cloud (wind from bottom right), with some iridescence in the center

Of the three principal high clouds, cirrus are the most familiar. Their icy wisps, including those known popularly as 'mares' tails' are often visible in the sky. They are the highest of the main cloud genera, and may sometimes even occur in the lower stratosphere.

Cirrus clouds generally occur as fine trails or filaments of cloud, which sometimes have a silken appearance. They may be fairly straight, curved, hooked, or apparently randomly entangled. One frequent form is the hooked shape, somewhat like a comma, at the top of which there is a small wispy tuft of cloud.

Parallel bands of cirrus are quite frequently seen, apparently radiating from one point of the sky. This type of structure is also often associated with jet streams, which are often made visible by the cirrus carried along by the high-speed, high-altitude winds. Cirrus billows are also commonly observed in jet streams.

On rare occasions, cirrus may appear as small tufts, generally with trailing wisps of ice crystals (virga), or even appear as tiny rounded heads or turrets that rise from a lower base. The small tufts often form in otherwise clear air.

Cirrus that is fairly high in the sky is always white, generally much brighter than any other clouds in same region of the sky. Dense patches do occur, however, that are thick enough to appear grey when seen against the light, and which may partially or even completely hide the Sun. Cirrus, being a high cloud, tends to show different colors to lower clouds. At sunset, it remains white when all lower clouds

CHARACTERISTIC FEATURES

- Fine threads or wisps of ice crystals, generally white, but appearing grey when dense and seen against the light.
- No precipitation at the ground.
- Frequently exhibits optical phenomena.

Cirrus uncinus trails, some with distinct generating heads

have taken on yellowish, orange, or red hues. By the time cirrus assumes these colors, the Sun has set below the horizon, and lower clouds often stand out as dark silhouettes against the dramatically lit high cloud. The effect is often even more striking at sunrise, when the air tends to be clearer.

Certain halo phenomena are frequent, particularly parhelia (mock suns), which are often stronger than those that appear in cirrostratus. Because cirrus tends to occur in patches, the larger optical phenomena (such as the 22° and 46° haloes) rarely appear in their entirety.

Although cirrus often forms in a clear sky, it may also arise from the virga (fallstreaks) that are a common feature of cirrocumulus and altocumulus, or through the decay of the thinner portions of a sheet of cirrostratus. Another major source is the cirrus plumes produced by cumulonimbus clouds, which often persist long after the parent clouds have dissipated. Such cirrus often trails behind the cold front of a depression.

A common method of formation of cirrus nowadays is from the contrails (condensation trails) of aircraft. When the conditions are appropriate, instead of dissipating, contrails may persist for hours, spreading into broad bands of cirrus that mask a considerable portion of the sky.

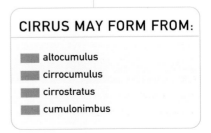

Dense cirrus spissatus during a spell of thundery weather

Distinguishing cirrus from other cloud types

Cirrus is most likely to be confused with cirrostratus, especially in the distance, when it may be difficult to see whether the cloud consists of individual trails or narrow bands (cirrus) or a fairly even layer (cirrostratus). Cirrus always occurs in relatively small patches or bands. It may also be difficult to decide

CIRRUS MAY FORM FROM:

- altocumulus
- cirrocumulus
- cirrostratus
- cumulonimbus

whether certain rounded tufts of cloud should be described as cirrus or cirrocumulus. Generally cirrocumulus will show a regular patterning of elements. Rounded cirriform heads and turrets are classed as cirrus (not cirrocumulus) when their apparent width is greater than 1°, measured 30° or more above the horizon.

One cirrus species (cirrus castellanus) has rounded heads or turrets and might be confused with the corresponding altocumulus species. It may, however, be distinguished from the latter by the distinct silken sheen that cirrus clouds exhibit. Dense cirrus could be mistaken for small patches of altostratus, but always appears white by comparison.

The only major cloud type that develops from cirrus is cirrostratus. This frequently occurs ahead of an approaching depression when the cirrus spreads out (almost imperceptibly) into a thin layer of cirrostratus.

CIRRUS SPECIES

There are five distinct species of cirrus:

- CIRRUS FIBRATUS (Ci fib) — FIBROUS CIRRUS
- CIRRUS UNCINUS (Ci unc) — HOOKED CIRRUS
- CIRRUS SPISSATUS (Ci spi) — DENSE CIRRUS
- CIRRUS CASTELLANUS (Ci cas) — CIRRUS WITH HEADS OR TURRETS
- CIRRUS FLOCCUS (Ci flo) — TUFTED CIRRUS

Cirrus fibratus (Ci fib)

Cirrus fibratus occurs as long, relatively straight or slightly curved streamers, which neither originate in tufts of cloud, nor show distinct hooks.

Cirrus fibratus

Cirrus uncinus (Ci unc)

Cirrus uncinus, on the other hand, show very distinct hooks at the leading ends of the trails.

Cirrus spissatus (Ci spi)

Cirrus spissatus is the dense cirrus that will partly or completely hide the Sun (or Moon) and which appears dark grey when seen against the light. Although it arises under various circumstances, it is particularly commonly found in the plumes or anvils of cumulonimbus clouds.

Cirrus castellanus (Ci cas)

Cirrus castellanus exhibits distinct turrets rising from a lower base of cloud. These turrets frequently appear to be arranged in lines, and the height of some of the them is greater than their width.

Cirrus floccus (Ci flo)

Cirrus floccus consists of small tufts of cloud, with ragged bases, often streaming away as virga.

Cirrus uncinus (TOP)
Cirrus spissatus forming the plumed top of a cumulonimbus cloud (MIDDLE)
Cirrus floccus (BOTTOM)

CIRRUS VARIETIES

Cirrus duplicatus radiatus (BELOW); irregular tangles of cirrus intortus (BOTTOM)

There are four varieties of cirrus:

- CIRRUS DUPLICATUS (Ci du) — MULTIPLE LEVELS OF CIRRUS
- CIRRUS INTORTUS (Ci in) — TANGLED CIRRUS
- CIRRUS RADIATUS (Ci ra) — CIRRUS IN EXTENSIVE BANDS
- CIRRUS VERTEBRATUS (Ci ve) — CIRRUS ARRANGED LIKE RIBS OR FISH BONES

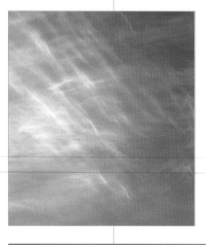

Cirrus, in common with all the stratiform cloud types may occur at two or more distinct levels, when it is classed as cirrus duplicatus. Similarly, the term cirrus radiatus is used to describe parallel bands of cirrus that stretch across the sky, and which appear to radiate (or converge) at one point on the horizon.

The intortus variety is confined to cirrus. Cirrus intortus appears as seemingly randomly entangled strands of cirrus with no recognizable pattern.

Finally, cirrus vertebratus shows a pattern that may be likened to that of the ribs or vertebrae of a skeleton, or else to fish bones. Again, this variety is exclusive to cirrus.

Cirrus radiatus clouds that have formed within the jet stream (TOP)

Although rather large, this formation is best described as cirrus vertebratus (BOTTOM)

Cirrostratus is a high cloud that is often overlooked, especially by those unfamiliar with the sky. Yet it displays some of the most striking halo phenomena, and is also a significant early indicator of a possible deterioration in the weather.

CHARACTERISTIC FEATURES

- A thin sheet of ice-crystal cloud, often striated, which frequently exhibits halo phenomena.
- No precipitation at the ground.

Typical cirrostratus nebulosus with a 22° halo

The general public may be aware of most of the ten major cloud types (even if they are unable to identify them), but cirrostratus often goes completely unnoticed. A veil of high cloud frequently steals across the sky and it is only when the sunshine seems to have lost some of its heat that people realize that thin cloud has invaded the sky. Cirrostratus is sometimes completely smooth in appearance, and only reveals its presence when a blue sky becomes slightly milky in appearance. More frequently, however, it shows a fibrous structure.

The edge of a layer of cirrostratus may be quite sharp, but quite frequently initial isolated wisps of cirrus gradually become more numerous

until they form a sheet of cirrostratus partially or completely covering the sky. This commonly occurs with the approach of the warm front in a depression system, and the cirrostratus itself gradually thickens, lowers towards the surface, and eventually turns into altostratus. Cirrostratus itself, however, is always thin enough for the Sun to cast shadows.

The most striking feature of cirrostratus is that it exhibits a whole range of halo phenomena: 22° and 46° haloes, parhelia (mock suns), and various other arcs and points of light. These effects (which are described in detail later) are so common that they are a good diagnostic for the pres-

ence of cirrostratus, but again they are generally overlooked. The optical phenomena are most marked when the cirrostratus is thin and gradually disappear as it thickens.

Cirrus fibratus thickening into cirrostratus

The main method of formation is, as just mentioned, the gradual ascent of warm air and the merging of streaks of cirrus. A sheet of cirrostratus may also arise from the ice crystals that fall from cirrocumulus clouds, or by the spreading out of the cirrus plumes at the top of cumulonimbus clouds. Less frequently it remains behind when altostratus clouds decay.

Distinguishing cirrostratus from other cloud types

Towards the horizon, cirrostratus may be difficult to distinguish from cirrus, but when overhead, its thin, extensive veil over a large part of the sky is quite unlike the patches and streaks of cirrus. It shows none of the general ripples, patches, rolls, or rounded masses that are found in cirrocumulus and altostratus. The presence of halo phenomena immediately distinguish cirrostratus from altostratus, and the latter is always a thicker cloud. Towards the horizon, however, the differences may becomes difficult to detect. Here the only guide is that both the general motion of the cloud layer and any changes occur more slowly in cirrostratus than in altostratus or even the much lower stratus.

In general, stratus is unlikely to be confused with cirrostratus, because of their considerable difference in heights, but if there is any doubt, the presence of any halo effects will confirm that the cloud is cirrostratus, because such phenomena are extremely rare in stratus, and can only occur under exceptionally cold conditions. Although stratus may sometime appear fibrous, the streaks are larger and much darker than those found in cirrostratus.

CIRROSTRATUS SPECIES

There are just two species of cirrostratus:

    CIRROSTRATUS FIBRATUS (Cs fib) – FIBROUS CIRROSTRATUS

    CIRROSTRATUS NEBULOSUS (Cs neb) – UNIFORM CIRROSTRATUS

An extensive sheet of cirrostratus fibratus

Generally a veil of cirrostratus will predominantly consist or one or other of these species. It either shows a fibrous nature (cirrostratus fibratus) or is completely featureless and uniform (cirrostratus nebulosus).

CIRROSTRATUS VARIETIES

Similarly, there are two varieties of cirrostratus:

- CIRROSTRATUS DUPLICATUS (Cs du) – MULTIPLE CIRROSTRATUS
- CIRROSTRATUS UNDULATUS (Cs un) – UNDULATING CIRROSTRATUS

The classification of cirrostratus duplicatus is used when more than one layer is present at any one time, although these may be difficult to detect, especially with cirrostratus nebulosus. The different layers become most apparent around sunset and sunrise, when the changing illumination, and consequent alterations in color and brightness of the layers helps to differentiate between them.

A layer of cirrostratus undulatus with a well-defined edge

The undulations in cirrostratus undulatus may be difficult to detect in the middle of the day, or when the cloud is directly overhead. They become more apparent towards the horizon, and may become very striking around sunrise or sunset, when the grazing illumination throws the crests and troughs into sharply defined relief. Under such conditions cirrostratus then bears some resemblance to the much higher and more finely structured noctilucent clouds that are occasionally visible in the middle of summer nights.

SEE ALSO
noctilucent clouds
(p. 82)

CIRROCUMULUS (Cc)

Rather like cirrostratus, the last of the high clouds, cirrocumulus, is often inconspicuous and overlooked, both because it is thin, and also because the individual cloud elements are small.

CHARACTERISTIC FEATURES

- A high-level layer cloud, broken into individual elements that appear white or pale blue, with no shading.
- There may be optical phenomena, but there is no precipitation.

CIRROCUMULUS MAY FORM IN CLEAR AIR OR FROM:

- cirrus
- cirrostratus
- altocumulus

Cirrocumulus is a very thin, white or pale blue cloud that occurs in patches or larger layers, and which consists of small units in the form of rounded masses and ripples, which may be separate or partially merged. Most of these elements are less than one degree across, measured 30° or more above the horizon, and may be more or less regularly arranged.

Cirrocumulus is generally far less distinctive than the similar lower clouds, altocumulus and stratocumulus. This is not only because, being higher, the individual cloud elements appear smaller, but also because it is much thinner than its lower counterparts. (There is, in fact, a direct relationship among these three types: the lower the cloud, the thicker the cloud layer and cloud elements.) Cirrocumulus is frequently so thin

Cirrocumulus showing various sizes of element and thickening into cirrostratus below left

that it appears merely as delicate ripples of cloud, which are so low in contrast that they are difficult to distinguish from the blue sky.

Cirrocumulus consists primarily of ice crystals, with some supercooled water droplets, although these tend to be converted into ice very rapidly. Being so thin and transparent, cirrocumulus always allows the position of the Sun and Moon to be readily detected, and objects on the ground invariably cast shadows. The even-sized cloud particles may produce optical phenomena in the form of coronae and iridescence from the light of both bodies.

Although the normal form of cirrocumulus is as a patch, sheet or layer, it may occur as smooth lens- or almond-shaped masses, and, very rarely, in the form of small tufts of cloud with ragged bases or as small turrets rising from a lower base.

Cirrus occasionally will transform into cirrocumulus, but the latter is more likely to arise from a layer of cirrostratus, in which shallow convection begins to occur, breaking up the layer. The decay of altocumulus may sometimes leave behind a patch of cirrocumulus. The smooth, lens-shaped species (cirrocumulus lenticularis) is caused by wave motion of the air.

Distinguishing cirrocumulus from other cloud types

Small heaps of cirrocumulus, sometimes in the form of rounded masses that rise from a more extensive base, may resemble similar tufts of cirrus. The latter, however, when they occur at least 30° above the horizon, always have an apparent diameter of more than one degree.

Unlike cirrus or cirrostratus, cirrocumulus is always broken up into tiny cloudlets and ripples. Most of the cloud consists of these, although it may also contain some parts that appear fibrous or smooth. These last two features are, however, more characteristic of cirrus or cirrostratus.

Perhaps the most common cause of doubt is the similarity between cirrocumulus and altocumulus. Again, the size of the cloudlets is all-important. If larger than one degree across, the cloud is altocumulus. Similarly, if the individual cloud masses have darker shading, then they are altocumulus. Cirrocumulus elements normally show no shading, and are often very low in contrast and difficult to see against the sky.

Cirrocumulus
stratiformis

SEE ALSO

altocumulus (p. 46)

cirrus (p. 54)

cirrostratus (p. 60)

corona (p. 104)

iridescence (p. 108)

supercooling (p. 150)

CIRROCUMULUS MAY BE CONFUSED WITH:

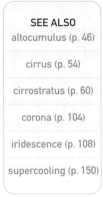

- cirrus
- cirrostratus
- altocumulus

CIRROCUMULUS SPECIES

There are four species of cirrocumulus:

- CIRROCUMULUS CASTELLANUS (Cc cas) — TURRETED CIRROCUMULUS
- CIRROCUMULUS FLOCCUS (Cc flo) — TUFTED CIRROCUMULUS
- CIRROCUMULUS LENTICULARIS (Cc len) — LENTICULAR CIRROCUMULUS
- CIRROCUMULUS STRATIFORMIS (Cc str) — EXTENSIVE SHEET OF CIRROCUMULUS

Cirrocumulus floccus with virga (TOP); cirrostratus stratiformis (BOTTOM)

Like the corresponding species of stratocumulus and altocumulus, cirrocumulus castellanus shows rounded turrets rising from a lower sheet or line of cloud. These towers may be difficult to see when the cloud is high overhead (partly because of the lack of shading), but are more readily visible in distant portions of the cloud cover. Again, as with the lower species, they are an indication of instability at cloud level.

Cirrocumulus floccus consists of tiny tufts of cloud, with rounded heads and ragged bases, which may sometimes show distinct virga. As with the lower altocumulus floccus, they may appear in large numbers spread across the sky, although their more tenuous nature means that they never present such a striking appearance.

Cirrocumulus lenticularis are smooth, lens- or almond-shaped clouds that form at the crests of otherwise invisible waves in the atmosphere. The clouds may be very long and normally have well-defined boundaries. Like the corresponding lower species (altocumulus lenticularis and stratocumulus lenticularis) they form when a stable layer of air is forced upwards, generally by mountainous land beneath, although they may sometimes occur far from any mountains. Cirrocumulus lenticularis, like most cirriform clouds, are often very tenuous. Because they consist of regularly sized particles they may also display iridescence under certain lighting conditions, although this is rarely as strong as in the similar, but higher, nacreous clouds.

The final variety, cirrocumulus stratiformis, is an extensive sheet of tiny cirrocumulus cloudlets that covers a substantial part of the sky. Although always thin, when light from the rising or setting Sun shines directly along the layer, signs of shading, absent at other times, may become apparent.

CIRROCUMULUS VARIETIES

Two varieties of cirrocumulus may sometimes be encountered.

- CIRROCUMULUS UNDULATUS (Cc un) – UNDULATING CIRROCUMULUS
- CIRROCUMULUS LACUNOSUS (Cc la) – CIRROCUMULUS WITH LARGE CLEAR HOLES

In cirrocumulus undulatus, the heaps of cloud may be arranged in rows (or billows), which have clear gaps between them. The individual cloudlets may be approximately circular, or may themselves be elongated parallel to the rows.

Cirrocumulus lacunosus is relatively rare, and consists of a layer of cloud with approximately circular holes, giving a net-like appearance. It does not often cover a large area of sky, but small patches occur quite frequently, surrounded by other cirriform cloud.

SEE ALSO

altocumulus castellanus (p. 49)

altocumulus floccus (p. 50)

altocumulus lenticularis (p. 50)

nacreous clouds (p. 80)

stratocumulus castellanus (p. 36)

stratocumulus lenticularis (p. 36)

virga (p. 78)

wave clouds (p. 94)

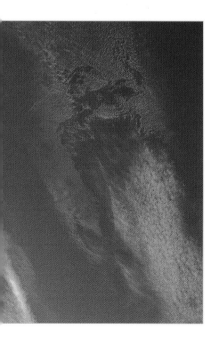

Supplementary features

Cirrocumulus is frequently accompanied by fallstreaks (virga), which consist of trails of ice crystals falling from the tufts of cloud. In general, these are not particularly long, but under rare conditions, when there is little wind shear in a deep cold layer of air, immensely long virga have been observed originating from apparently insignificant tiny cloudlets.

Mamma (downward bulges or pouches) are sometimes seen, but these are only rarely well-developed, unlike those seen in lower clouds.

A thin layer of cirrocumulus lacunosus

Cumulonimbus are the giants among clouds. Their bases may be close to the ground, but their tops tower right up in the atmosphere. They often reach right through the troposphere (the lowest layer of the atmosphere), in which most weather occurs. Even when relatively small, they may bring heavy downpours of rain, while larger systems give rise to thunder and lightning, hail, violent winds, and even, in the largest systems, destructive tornadoes.

CHARACTERISTIC FEATURES

- A massive, deep cloud, with a dark base and brilliantly white top when fully illuminated.
- The top may be rounded, but starting to lose its hard appearance, or may have become ragged in the form of a cirrus plume, or flat anvil.
- Heavy precipitation may be present, as well as lightning and thunder.

Most cumulonimbus clouds are immense, towering high into the sky and extending over wide areas of countryside. Frequently, in fact, their horizontal extend is so great that the clouds' characteristic features are difficult to appreciate unless one is at a considerable distance and has a clear view. They are always extremely dense clouds that appear brilliant white when fully illuminated by the Sun, but which, by contrast, seem exceptionally dark grey or even black when viewed against the sunlight. Their bases, in particular, are always extremely dark and ragged with falling rain, hail or, under certain conditions, snow. This precipitation is often in the form of virga (fallstreaks).

Cumulonimbus clouds sometimes reach throughout all three conventional levels. Their tops often consist of multiple, rounded, rising heads of cloud. Normally some of these—the lower ones, or those that are growing particularly rapidly—appear hard and sharp against the sky. Some of the highest cloud towers lose their hard outline and show a slightly softer appearance, which generally soon becomes fibrous. These features, together with the presence of precipitation beneath the cloud, positively identify the cloud as a cumulonimbus. They mark this type of cloud's two species: cumulonimbus calvus and cumulonimbus capillatus, described in more detail shortly.

Once the top of the cloud becomes fibrous, it may also spread out into another highly characteristic form, that of a flattened anvil (cumulonimbus incus), or appear as a plume of cirrus, which may be rapidly drawn out across the sky by high-altitude winds.

Cumulonimbus incus with mamma

The correct identification of cumulonimbus is important, and should be taken as a warning of potentially severe weather, particularly by those engaged in outdoor pursuits. Not only do these clouds give heavy precipitation, including damaging hail, but they produce violent gusts and changes of wind direction, as well as lightning and other hazards.

Although under certain circumstances, cumulonimbus clouds may be relatively small and of limited horizontal extent, sometimes consisting of just a single active cell, frequently several cells in various stages of growth may be clustered together in a single system. Both types are often described by meteorologists and weather forecasters as 'showers'. On other occasions, cumulonimbus may build up into massive walls of cloud or more complex and violent systems, known as squall lines, and multicell or supercell storms. The last of these may spawn highly destructive tornadoes.

A cluster of cumulonimbus cells, the most distant of which has spread into an anvil

SEE ALSO

cumulus congestus (p. 24)

multicell storms (p. 169)

nimbostratus (p. 44)

showers (p. 164)

supercell storms (p. 169)

tornado (p. 174)

virga (p. 78)

Distinguishing cumulonimbus from other cloud types

Cumulonimbus clouds obviously have much in common with large cumulus congestus clouds. As soon as part of the cloud-top loses its hard outline and becomes fibrous or striated, the cloud may be classified as cumulonimbus. If the upper part of the cloud is invisible, however, the presence of hail, lightning or thunder serve to confirm that the cloud is a cumulonimbus.

Extensive cumulonimbus may sometimes be difficult to distinguish from nimbostratus, again especially if the tops of the cloud cannot be seen. Nimbostratus, however, is generally associated with different conditions and frequently develops gradually from altostratus. It also produces more continuous precipitation (albeit sometimes with heavy bursts), rather than the shorter, intense showers characteristic of cumulonimbus. Again, by convention, if there is hail, lightning or thunder, the cloud is classified as cumulonimbus.

CUMULONIMBUS SPECIES

SEE ALSO

cold rain (p. 150)

jet stream (p. 98)

precipitation (p. 146)

warm rain (p. 150)

There are just two cumulonimbus species:

 CUMULONIMBUS CALVUS (Cb cal) — SMOOTH ('BALD') CUMULONIMBUS

CUMULONIMBUS CAPILLATUS (Cb cap) — FIBROUS CUMULONIMBUS

These two species are only found in cumulonimbus, so recognition of either of them automatically means that the cloud is a cumulonimbus.

Cumulonimbus calvus (Cb cal)

In cumulonimbus calvus, although calvus means 'bald' in Latin, to most people this probably conjures up the wrong image. The term calvus is applied when one of the rising cloud towers loses its hard outline and takes on a slightly softer form. This change occurs when the water droplets in the cloud freeze into ice crystals. This is actually a very important process in the formation of rain.

(FAR RIGHT)
Cumulonimbus with cells at the calvus, capillatus and incus stages

Cumulonimbus capillatus (Cb cap)

As more and more ice crystals form in the uppermost layers of the cloud, this gradually changes from the soft, calvus form and takes on a fibrous appearance, when it is known as cumulonimbus capillatus (from the Latin word for 'hairy'). Depending on the exact wind conditions at height, the ice crystals may fall more or less vertically, often giving rise to distinct virga (fall-streaks). At other times the crystals may be carried horizontally and spread out into an overhanging anvil. On yet other occasions, if the winds are very strong—such as when cumulonimbus towers reach the jet stream—a vast cirrus plume may form, which is rapidly carried downwind.

There are no cloud varieties associated with cumulonimbus but, by way of compensation there are three accessory clouds and several supplementary features, all of which are described more fully later (pp. 72-79). Although none occur exclusively with cumulonimbus, many are most commonly associated with this cloud type. Cumulonimbus is, in fact, the only type in which all might occur simultaneously.

Accessory clouds

PANNUS: ragged clouds beneath the main cloud base

PILEUS: a thin cap of cloud above rising towers

VELUM: a thin layer of cloud separate from the main cloud mass

Supplementary features

ARCUS: an arch or roll of cloud

INCUS: a flattened anvil shape

MAMMA: pouches of cloud, often below an anvil

PRAECIPITATIO: rain, snow or hail reaching the ground

TUBA: a funnel cloud

VIRGA: precipitation that does not reach the ground.

Cumulus congestus (TOP)
Cumulonimbus calvus (MIDDLE)
Cumulonimbus capillatus (BOTTOM)

ACCESSORY CLOUDS

Accessory clouds are always found in association with one of the main cloud types, never on their own.

PANNUS (pan)

When precipitation falls from a cloud into clear air beneath, it tends to evaporate, cooling the layer and causing its humidity to increase. Turbulence within this lower layer may cause some parcels of air to rise and reach their condensation point, producing broken fragments of cloud beneath the main cloud layer. If the precipitation continues, these ragged fragments (a form of stratus fractus) may become quite extensive,

Ragged pannus often occurs beneath cumulonimbus and nimbostratus clouds

tending to merge with the cloud above. This type of accessory cloud is frequently found beneath nimbostratus and cumulonimbus, as well as associated with precipitating cumulus congestus and altostratus, although in the last case, being higher, it is often more difficult to make out distinctly.

Pileus forming a temporary smooth cap over a rising cumulus congestus cloud

PILEUS (pil)

The accessory cloud known as pileus (from the Latin word for 'cap') occurs when a vigorously rising cumuliform cloud (usually cumulus congestus or cumulonimbus) encounters a hitherto invisible humid layer. The rising bubble of air causes a localized lifting of the humid layer and, if conditions are just right, the air reaches its dewpoint, and produces a hood or cap of cloud above the rising tower. Such clouds are very similar in appearance to lenticular clouds—and arise through a somewhat similar mechanism—but, unlike the wave clouds, pileus are normally very short-lived. The vigorous cumulus or cumulonimbus generally continues rising and penetrates the humid layer, which then looks like a collar around the main tower. The latter usually grows sideways as it rises and the mixing within it and entrainment of the surrounding air quickly incorporates the pileus into the main cloud. This short life-time is quite unlike the persistence shown by the majority of lenticular clouds.

Very occasionally, there may be several shallow, humid layers one above the other, so that multiple pileus caps and collars may occur, again somewhat similar to the stack or 'pile d'assiettes' seen with wave clouds.

VELUM (vel)

SEE ALSO

altostratus (p. 40)

cumulonimbus (p. 68)

cumulus congestus (p. 24)

nimbostratus (p. 44)

pile d'assiettes (p. 95)

stratus fractus (p. 30)

thermals (p. 86)

wave clouds (p. 94)

Velum is related to pileus, but in this case the whole layer is visible, forming a thin, horizontal sheet or veil lying above the top of cumuliform clouds, and frequently penetrated by the tallest cloud towers. There is often little sign of entrainment, and the veil may persist for an extremely long time, frequently remaining visible well after convection has begun to die down and the cumuliform clouds have started to decay.

Two layers of velum, left behind by shower clouds, are visible here

SUPPLEMENTARY FEATURES

There is obviously a vast range of different forms that clouds may assume, but six are caused by specific processes and are relatively consistent in appearance, so they have been adopted as part of the World Meteorological Organization's official cloud classification. Many other forms do have well-established and widely accepted names, and some of these will be described later when discussing particular systems and atmospheric processes.

ARCUS (arc)

A dense roll of cloud of considerable horizontal extent, found on the lower, leading edge of an active cloud system. It most often occurs in association with cumulonimbus clouds or the larger multicell and super-cell systems, but may occasionally be seen with less vigorous cumulus clouds. The edges are often more-or-less ragged and when the cloud is particularly heavy and extensive it appears as a dark arch across the sky. (It is also known as an arch cloud, from which it derives its name.) Both shelf clouds and the rarer roll clouds, described later, are forms of arcus.

Arcus: the leading edge of the gust front from a large cumulonimbus cloud over Sydney Harbour

INCUS (inc)

A highly characteristic 'anvil' form that occurs only with cumulonimbus clouds (cumulonimbus incus). The vigorous cells in a cumulonimbus cloud often cease to rise only when they encounter a major inversion, where temperature begins to increase with height. This is often at the level known as the tropopause, which separates the lowest layer of the atmosphere (the troposphere) from the overlying stratosphere. The enormous numbers of ice crystals that form are born away by the generally faster winds at this level, giving rise to a flattened shield of cirrus that stretches downwind. In most cases, the circulation in the rising cells is so strong that, on reaching the inversion, some of the cloud expands upwind, but to a lesser extent than downwind. This produces the classic anvil shape, with a characteristic small overhanging shelf of cloud on the upwind side, and a much larger one downwind.

SEE ALSO

roll cloud (p. 167)

shelf cloud (p. 167)

cumulonimbus (p. 68)

supercell storms (p. 169)

multicell storms (p. 169)

A pair of cumulonimbus incus clouds in winter, when the tropopause was relatively low

MAMMA (mam)

This cloud form consists of bulges or pouches on the underside of a particular cloud. (It is named from the Latin word for 'udder'.) These features may be found on several types of cloud: cirrus, cirrocumulus, altocumulus, altostratus, stratocumulus, and cumulonimbus. They arise when a localized downdraft carries cold air down into a warmer layer, causing it to cool to its dewpoint and produce cloud droplets. This process has been termed 'upside-down convection', drawing a parallel with the commonly seen process where a thermal carries warm air upwards and gives rise to cloud.

Dramatically lit mamma beneath a cumulonimbus anvil, photographed in Poland

The mamma seen beneath many types of cloud are relatively small and not particularly pronounced, but beneath cumulonimbus clouds they may assume various forms and are sometimes extremely striking. Beneath the main body of the cloud, they may appear as long, distorted tubes, but underneath a cumulonimbus anvil, in particular, they often appear as enormous globular masses, which are especially striking when illuminated by low sunlight. Here they arise because the top of the anvil loses heat to space, creating the cold downcurrents. The bases of such

Long, distorted
mamma beneath
a cumulonimbus
cloud

SEE ALSO
cumulus congestus
(p. 44)

cumulonimbus
(p. 68)

precipitation (p. 146)

mamma may persist even after the main cumulonimbus cloud and the cirrus anvil have disappeared.

PRAECIPITATIO

This is the term used when any cloud is producing precipitation in any form (liquid or solid), that is reaching the ground. It covers everything from drizzle or a light dusting of snow to torrential rain and destructive hail. It most commonly applies to nimbostratus, cumulus congestus, and cumulonimbus, but — depending on circumstances — may also be encountered with altostratus, stratocumulus, and stratus. It is closely related to the next feature, virga.

Praecipitatio:
a heavy
convective shower,
Gemsbokvlaate,
Namibia

VIRGA

Otherwise known as 'fallstreaks', these are trails of precipitation that are observed to fall from a cloud, but which evaporate before reaching the ground. They occur with a wide range of clouds: cirrocumulus, altocumulus, altostratus, nimbostratus, stratocumulus, cumulus, and cumulonimbus. Depending on the conditions in the layer into which the particles are falling, virga may present various different forms. Although not normally regarded as such, cirrus clouds could be considered to consist solely of virga, because they comprise a generating head (although this is not always readily visible) and a trail of ice crystals. On occasions, when conditions are uniform throughout a considerable depth, the virga may be almost perfectly vertical fallstreaks. Yet if there is considerable wind-shear they may be drawn out into almost horizontal trails. Inexperienced observers sometimes mistake dense, almost horizontal virga for the shadows of condensation trails.

Well-defined virga beneath altocumulus clouds consisting of ice crystals melting into water droplets

Associated with virga are 'fallstreak holes': openings that occur in a layer of altocumulus or cirrocumulus, with virga beneath them. These holes are often highly symmetrical, and some display an extraordinary degree of regularity, appearing as almost perfectly circular holes in the cloud layer. They occur when glaciation (freezing) begins at a particular point in the cloud and spreads out more-or-less evenly, but the precise conditions that allow this to occur are poorly understood. On occasions, the virga themselves may dissipate, leaving a symmetrical hole in the cloud as the only sign of what has happened. A similar process may sometimes be initiated by an aircraft, leaving a long clear lane in the cloud, with virga beneath it.

TUBA

The term tuba is applied to any column or cone of cloud that extends down from the main cloud mass towards the surface. (It thus differs from devils and whirls, which extend upwards from the surface, and which are described later.) Such funnel clouds that do not reach the ground are quite common: far more common than the general public realizes and, despite the prominence they are sometimes given in the media, are not particularly noteworthy from a meteorological point of view. They occur beneath vigorous cumulus congestus and cumulonimbus clouds (in particular), along gust fronts, and in association with multicell, supercell, and other major storm systems (such as tropical cyclones).

Only when such a funnel cloud touches down is it classed as a landspout, waterspout, or true tornado. The last of these forms through a different mechanism than the others, but all will be described in detail later.

SEE ALSO

contrails (p. 100)

cumulus congestus (p. 44)

cumulonimbus (p. 68)

distrails (p. 101)

landspouts (p. 173)

tornadoes (p. 174)

waterspouts (p. 173)

whirls (p. 172)

A typical funnel cloud, or tuba, which later touched down and became a waterspout (p. 175)

NACREOUS CLOUDS

CHARACTERISTIC FEATURES

- Beautifully colored, high-altitude clouds very similar to cirrocumulus lenticularis, usually seen against a fairly dark sky shortly after sunset or before sunrise.
- A white form is occasionally observed.

From time to time, at fairly high latitudes (generally greater than 50°N or S), there are striking displays of beautiful, multi-colored clouds, visible just after sunset or before sunrise. At these times the sky itself is fairly dark, but because of their altitude nacreous clouds are illuminated by the Sun when lower clouds are in darkness. The pastel colors seen in these displays are so different from those normally visible in clouds, that the events usually provoke widespread interest and comment on radio and television and in the newspapers. The early-morning displays are less-widely noticed, because fewer people are awake before dawn.

Delicate pastel shades are visible in these nacreous clouds seen at sunset

The colors in these nacreous clouds (also known as mother-of-pearl clouds), are actually a form of iridescence, similar to that seen on the edges of certain clouds during the daytime. The colors arise through the phenomenon known as diffraction, where rays of light are deviated from their original paths by the cloud particles. The wavelength that is diffracted (and thus the apparent color) is strongly dependent on the size of the particles. The colors are particularly pure and strong in nacreous clouds because different parts of the clouds contain large numbers of particles that are identical in size, thus producing bands of pure color.

The white form of nacreous clouds was here photographed over Antarctica

SEE ALSO:
cirrocumulus lenticularis (p. 66)

iridescence (p.108)

stratosphere (p. 188)

wave clouds (p. 94)

When the Sun is low, nacreous clouds take on sunset colors

Nacreous clouds belong to the class known to meteorologists as polar stratospheric clouds, which form at altitudes of 9–18 mi. Two slightly different types of cloud occur. Brilliantly colored nacreous clouds are created when there is wave motion at their altitude, caused by the airflow over mountains beneath. (They are high-altitude wave clouds and closely resemble cirrocumulus lenticularis.) The air rises rapidly and is cooled abruptly, producing large numbers of small, evenly sized particles of ice—and thus brightly colored clouds. These are the clouds most commonly seen in the northern hemisphere. The other type of cloud is also a wave cloud, but occurs when persistent winds are accompanied by a slow drop in temperature, as with the onset of the winter season. The ice particles produced are fewer in number, but much larger in size. Because of their size, they diffract light over a whole range of wavelengths, so the clouds shine with a white light, sometimes with a hint of color around the edges. This type appears to be more commonly seen in the Antarctic than in the Arctic.

All polar stratospheric clouds (including a third type not discussed here) are involved in the destruction of ozone that has led to the formation of ozone holes in both polar regions. (The cloud particles provide the necessary sites for the chemical reactions that destroy ozone.) Nacreous clouds appear to be becoming more frequent in the northern hemisphere. In the past displays were seen every 30–40 years, but several major displays were reported in the 1990s alone, when the northern ozone hole first became apparent.

NOCTILUCENT CLOUDS (NLC)

The vast majority of clouds—even nacreous clouds, which occur in the stratosphere—are invisible at night, unless illuminated by the Moon. There is, however, one type of cloud that is sometimes visible around midnight. These noctilucent clouds—the name means 'shining at night'—are visible from certain parts of the Earth around the time of the summer solstice. They are relatively easy to photograph.

CHARACTERISTIC FEATURES

- Tenuous clouds with a characteristic bluish-white or yellowish coloration, visible in the direction of the pole in the hours around midnight.
- Their structure somewhat resembles certain species of cirrus and cirrostratus.

Noctilucent clouds are the highest clouds in the atmosphere. They occur slightly below the mesopause (which is the upper boundary of the mesosphere), at altitudes of 48–51 mi. This is so high that, for observers at high latitudes (beyond about 45° N or S), and for about a month before and after midsummer, the clouds are illuminated by the Sun while the observers are in darkness. The clouds themselves do not extend closer to the poles than about latitude 60°, although they are probably identical to mesospheric clouds known to exist over the poles. It is believed that a separate layer occurs over the equatorial regions, but here the clouds may be glimpsed only briefly at sunrise and sunset.

At high latitudes, of course, twilight persists throughout the night. Even so, the clouds are usually highly distinctive and readily recognizable, especially during a major display. Their appearance somewhat resembles that of cirrus or billows in cirrostratus. Around midnight, they have a characteristic bluish-white coloration, but earlier in the evening and later towards dawn they may appear tinged with yellow, when the sunlight has travelled through a greater length of atmosphere.

The clouds actually occur as a very thin layer, and may be quite invisible to observers located directly beneath them, but readily seen by others farther away from the display. The various structures arise mainly because the layer has undulations, caused by waves in the high-altitude easterly winds. The variations in apparent density occur because the number of cloud particles encountered along the line of sight differs from one point to another. The particles consist of ice, but there is no general agreement about the nature of the freezing nuclei, which may be particles of dust from meteors, or clusters of ions created by incoming cosmic rays. Nucleation actually occurs slightly higher than the visible cloud layer, at the mesopause, the coldest part of the atmosphere, where the temperature is about 150 K (−253°F). Because aurorae cause heating of the upper atmosphere, it was once thought that noctilucent clouds and aurorae could not occur at the same time. Some years ago, however, they were both captured in a single photograph, taken by an observer in Canada.

There are four main structural forms:

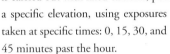 TYPE I: VEILS — TENUOUS FILMS, RESEMBLING CIRRUS OR CIRROSTRATUS, SOMETIMES WITH A SLIGHT FIBROUS STRUCTURE. OFTEN PRECEDE, OR OCCUR AS A BACKGROUND TO OTHER FORMS.

TYPE II: BANDS — LONG STREAKS, OFTEN IN GROUPS THAT ARE APPROXIMATELY PARALLEL, BUT SOMETIMES CROSSING AT A SHALLOW ANGLE. THEY MAY CHANGE IN BRIGHTNESS OVER PERIODS OF 20–60 MINUTES.

TYPE III: BILLOWS — CLOSELY SPACED, APPROXIMATELY PARALLEL, SHORT STREAKS. SOMETIMES APPEAR ON THEIR OWN, OR MAY CROSS THE LONGER BANDS. BILLOWS OFTEN CHANGE FORM AND PATTERN AND ALTER IN BRIGHTNESS WITHIN MINUTES.

TYPE IV: WHIRLS — PARTIAL OR COMPLETE LOOPS OR RINGS OF CLOUD WITH DARK CENTERS. CLOSED RINGS ARE RARE.

Various subdivisions and more complex forms are recognized by specialist observers. Bright knots may occur, for example, where two sets of bands cross one another.

The clouds are carried west-southwest (in the northern hemisphere) by the easterly winds, but the structures often move in the opposite direction. Their study provides information about upper-atmosphere winds, and this is carried out with fixed cameras, pointing in a particular direction and at a specific elevation, using exposures taken at specific times: 0, 15, 30, and 45 minutes past the hour.

Noctilucent clouds are unlikely to be confused with any other form of cloud, except the persistent trails left by some bright meteors, or the exhaust trails from rocket launches. Both of these tend to be single trails that are distorted by upper-atmosphere winds, and rocket trails often show strong iridescence and, in some respects more closely resemble nacreous clouds.

SEE ALSO

cirrostratus (p. 60)

cirrus (p. 54)

mesosphere (p. 188)

nacreous clouds (p. 80)

stratosphere (p. 188)

wave clouds (p. 94)

Characteristic silvery-white noctilucent clouds photographed around local midnight

AURORAE

Although aurorae are not clouds and they do not have a direct relationship with the weather, they are included here because they are seen during twilight and at night, like nacreous and noctilucent clouds. Some types of noctilucent clouds do resemble certain auroral forms.

Aurorae occur at both northern and southern latitudes, being known formally as aurora borealis and aurora australis (or the Northern and Southern Lights). They are produced when energetic particles from the Sun, which have penetrated the outer regions of the Earth's magnetic field, are accelerated along the magnetic-field lines and collide with atoms and molecules in the upper atmosphere at altitudes of about 60 mi. and above. The excited molecules emit highly characteristic colors, the most common being green from oxygen atoms, and red from nitrogen molecules. Sometimes, when the top of a display is in sunlight (at about 600 mi. altitude), there is a purple-violet color from nitrogen.

An early stage in the development of rayed structure in a red auroral band

The colors that are actually seen depend strongly on the observer's eyesight and the strength of the aurora. Weak aurorae often appear almost colorless, or a pale green. People who have poor low-light red sensitivity may be unable to see the red and purple-violet tints in even major displays, despite seeing the pale green coloration easily. Red aurorae are often mistaken for distant fires.

Aurorae are seen quite frequently from Northern Europe, Siberia and Canada (and also from Antarctic regions), but naturally tend to be invisible during the long summer twilight. Occasional major displays are visible anywhere on Earth, but except for these rare occasions, aurorae normally appear in the direction of the pole. There are several major forms:

ARC: AN ARCH-LIKE STRUCTURE, WITH A DISTINCT LOWER EDGE, BUT MORE DIFFUSE UPPER BORDER

BAND: A DEEP RIBBON-LIKE STRUCTURE, FOLDED LIKE CURTAINS; OFTEN WITH RAPID MOVEMENT

CORONA: A SERIES OF BROAD RAYS THAT APPEAR TO RADIATE FROM A POINT OVERHEAD

GLOW: A WEAK GLOW ON THE POLAR HORIZON

PATCH: LUMINOUS AREA RESEMBLING A CUMULUS CLOUD; SEEN LATE IN A DISPLAY AND OFTEN PULSATES IN STRENGTH

RAY: AN APPROXIMATELY VERTICAL STREAK OF LIGHT; FREQUENTLY SEEN WITH ARCS AND BANDS

VEIL: A FAINT, EVEN GLOW OF LIGHT ACROSS A LARGE PART OF THE SKY

SEE ALSO

nacreous clouds
(p. 80)

noctilucent clouds
(p. 82)

Displays may vary greatly in their strength and activity. Some may never be more than a weak glow or veil, which may be visible for just a short time or persist throughout the night. Larger displays often begin as a glow, which strengthens to become a distinct arc across the sky. Many displays never progress beyond this form, but frequently rays appear along the length of the arc. Such rays actually follow the lines of the Earth's magnetic field.

Rays may be relatively quiescent, but in other displays they vary rapidly in length, brightness and position. The arc often becomes a distinct band, which waves around like a curtain in the breeze. In strong displays, multiple rayed bands may also appear, and the whole display may expand towards the equator. If it passes overhead, the display takes on the form of a corona (which is quite unrelated to the optical phenomena with the same name). Later, towards dawn, the display may break up into pulsating patches of light.

Generally the lower part of a display is green, with red in the upper regions, but all-red aurorae are not uncommon. Yellow, blue, violet and white are also sometime reported, often appearing simultaneously in different parts of the display. The strength of individual aurora may vary in a highly distinctive fashion. These variations are known as:

PULSATIONS: SLOW FLUCTUATIONS WHERE THE BRIGHTNESS VARIES OVER A PERIOD OF MINUTES

FLICKERING: RAPID VARIATIONS AFFECTING PART OR ALL OF A DISPLAY

FLAMING: SURGES OF BRIGHTNESS THAT SWEEP UPWARDS FROM THE HORIZON

A homogeneous arc, showing the characteristic auroral green coloration

Displays frequently occur at both northern and southern latitudes simultaneously, although because of the seasonal difference, one display is normally more spectacular than the other.

Clouds form when air is cooled below its dewpoint. In the atmosphere, cooling occurs in just two principal ways: either by rising and expanding, or by coming into contact with a cold surface.

There are various processes which may cause air to rise in the atmosphere, some of which have been briefly mentioned in the earlier descriptions of individual cloud types. There are three main reasons why air may rise:

- THROUGH CONVECTION, GENERALLY CAUSED BY HEATING FROM BELOW
- THROUGH FORCED ASCENT OVER HILLS OR MOUNTAINS (OROGRAPHIC LIFTING)
- BY FRONTAL LIFTING, BEING UNDERCUT BY DENSER, COLDER AIR

There is also a fourth process, known as convergence, where air is forced into a restricted area, and must rise to escape.

Convective clouds

All the cumuliform clouds arise through convection, which is frequently encountered in everyday life. The process has been discussed under cumulus clouds, where thermals rise from a heated surface (normally the ground). Although often described as bubbles of warm air, there is actually a circulation in thermals, with an updraft in the center, and a weaker (because less concentrated) downdraft around the outside. As the thermal rises, it tends to expand, and the circulation mixes cooler air from the surroundings into it, which causes it to weaken and eventually decay. Even when a thermal reaches the condensation level, the circulation is preserved, and indeed the centers of the bases of cumulus clouds tend to be slightly higher than the circumference.

Convection may also be initiated when the top of a cloud is cooled by radiating heat away to space. This often occurs with a fairly even layer of stable stratiform cloud. The cooler air starts to sink and breaks up the layer into shallow, individual convective cells. This mechanism of 'upside down convection' is often responsible for the formation of stratocumulus, altocumulus, and cirrocumulus. A similar process occurs in the formation of mamma, which may be found beneath cumulonimbus clouds (particularly beneath anvils), and in stratocumulus and altocumulus layers that have been produced by the spreading out of cumulus clouds at an inversion. Humid air descends within the mamma themselves, and drier air ascends between them.

Orographic clouds

If the wind forces air over a mountain barrier, expanding and cooling as it does so, it may rise above the condensation level. The type of cloud that forms will depend upon the precise nature of the air, in particular on its stability or instability. When the air is stable, it will give rise to stratiform clouds, such as stratus or altostratus, and the peaks may become shrouded in cloud, which to people on the hills themselves appears as fog. With unstable air, cumuliform clouds will grow and, again depending on the exact conditions, these may vary from small cumulus to enormous cumulonimbus clouds. On occasions, the flow of air may produce long-lasting, and essentially stationary cumulonimbus clouds with prolonged heavy rainfall. Such orographic lifting is often the cause of flash flooding in mountainous areas.

Under certain circumstances, the amount of uplift and cooling may be sufficient to create instability in otherwise stable air. This may lead

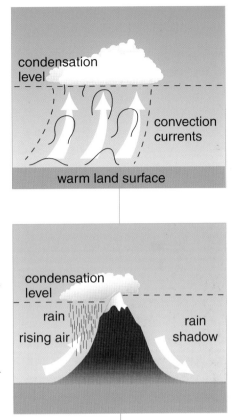

Convection (TOP)
Orographic cloud (ABOVE)

An inversion is limiting the growth of cumulus clouds and causing them to spread out into moderately thick stratocumulus

SEE ALSO

cumuliform
clouds (p. 14)

depressions (p.180)

fronts (p. 179)

stratiform
clouds (p. 14)

to the situation where there are stratiform clouds over the plains, but cumulus or even cumulonimbus over the hills.

Naturally, when the air descends to leeward of the hills or mountains, it is compressed and warms, so any clouds tend to evaporate. This is often seen over oceanic islands, which may be capped by cloud, but surrounded by clear skies. An interesting situation arises when large amounts of precipitation fall on the windward side of the hills, triggered by the uplift. When the air descends on the other side of the barrier, it actually warms at a greater rate, because moisture has been removed in the rain, sleet or snow. As a result, the temperature on the leeward side may be higher than at the same altitude to windward. These are known as föhn conditions and were first recognized in Europe, where warm, humid air from the south is often drawn north over the Alps. It has since been found that there is also a tendency for air to be drawn down from yet higher levels, giving an even greater warming effect. The Chinook that descends the eastern side of the Rockies is one example of such a föhn wind, the onset of which may rapidly thaw an existing snow cover.

Frontal clouds

Where different air masses meet, typically at a front, and in particular in depressions, the two bodies of air have different temperatures (and humidities) and the colder air forces the warm air away from the surface. The type of cloud that is formed depends greatly on the exact characteristics of the air masses concerned, and especially on the type of front. If the warm air is advancing on the cold (at a warm front), it slides up a gently inclined surface. Because warmer air is overriding colder air, producing an inversion and stable conditions, stratiform clouds (in particular cirrostratus, altostratus and nimbostratus) are created.

Where cold air is advancing on the warm air (at a cold front), the cold air vigorously undercuts the warm air and forces it to rise. The clouds are often similar to those at a cold front, but (particularly over continental areas) the fact that cold air is advancing over a warm surface frequently results in unstable conditions, leading to vigorous convection and cumuliform clouds, particularly cumulonimbus.

The structure of the fronts in depressions and their associated clouds will be discussed in more detail later.

Frontal cloud

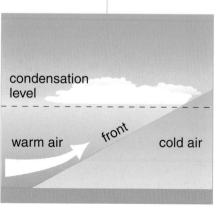

condensation level

warm air   front   cold air

An orographic cap cloud over a peak on Jan Mayen Island (ABOVE)

Cumulus mediocris clouds caused by uplift over Snowdonia (RIGHT)

BILLOWS

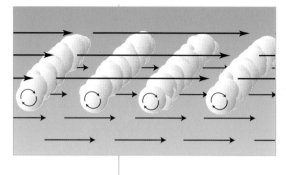

Clouds often form in regular patterns, and this is most marked in the parallel rolls of cloud seen in the undulatus varieties of certain cumuliform clouds: stratocumulus, altocumulus, and cirrocumulus. Somewhat similar undulations are also seen in stratus, altostratus, and cirrostratus.

Such billows, as they are called, stretch approximately at right-angles to the wind direction. In fact, they are generally created by wind shear, which occurs either when winds at different levels are in slightly different directions, or when they are of different strengths. (In many cases both conditions apply.) This variation in wind speed and direction creates a series of waves at that level. There are various forms these billows may take. The cloud may break up into a series of rolls, separated by clear air, or form with undulating surfaces. The top may be undulating and the bottom—at the condensation level—almost flat, with thicker portions (in the crests of the waves), and thinner cloud in the troughs. Large undulations may sometimes be seen on the top of a sheet of stratus cloud, when looking down onto it from a mountain or aircraft.

Billows also frequently occur when a stable layer of air is forced to rise,

The formation of billows (TOP)

A fine display of billows in altocumulus (altocumulus undulatus radiatus) over Dublin in the Republic of Ireland (ABOVE)

An unusual set of billows, restricted to a narrow line of waves, which rapidly disappeared (RIGHT)

Extremely short-lived billows on the top of a layer of sea fog (BELOW)

and this may create cloud formations that resemble the breaking waves observed on water. Cirrus may sometimes be seen in this form. Such waves are extremely short-lived, often lasting no more than one or two minutes, so you are lucky if you see them, and even luckier if you are able to capture them in a photograph. Such 'breaking' waves are sometimes seen on the upper surface of banks of mist and fog, but here again they are short-lived.

Billows may sometimes have a superficial resemblance to wave clouds, especially when there are wide gaps between the rolls of cloud. There is, however, one very distinct difference. Billows always move downwind, unlike wave clouds which remain stationary in the sky. Wave clouds may also be very much larger, and it is perfectly possible—and quite commonly observed—for wave clouds to exhibit a series of billows.

To complicate the matter even further, another mechanism may sometimes come into play, causing undulations know as corrugations to appear. These, unlike billows, are aligned parallel with the wind direction, and may be seen in both billow and wave clouds. In some clouds, corrugations are more prominent than billows, and they may occasionally be seen projecting like narrow fingers from the downwind edges of clouds (which is where the cloud droplets are evaporating).

SEE ALSO
fog (p. 142)

wave clouds (p. 94)

CLOUD STREETS

SEE ALSO

anticyclone (p. 182)

inversion (p. 187)

radiation fog (p. 142)

stratocumulus
(p. 32)

Cumulus clouds often form in a line, known as a cloud street, that runs downwind. Sometimes wide areas are covered by a whole series of parallel streets, with clear air between them.

Single cloud streets may be produced if there is a strong source of heating, which is able to create a constant supply of thermals that feed the individual cumulus clouds. Such cloud streets are quite commonly seen trailing downwind from oceanic islands. If the wind is blowing along the edge of a strongly heated area (such as the edge of an escarpment) it may create a street in just the same manner.

Multiple, parallel streets are a recognized variety of cumulus, cumulus radiatus. The most favorable conditions for their formation occur when the lowermost layer of air is unstable, but is capped by an inversion—by a stable layer of air. This often occurs when upper air is subsiding, such as under anticyclonic conditions, and is also frequently found when radiation fog has formed overnight. Convection occurs below the inversion, with air rising in thermals below the clouds and sinking more gently in the clear air between the streets. The spacing between the lines of cloud is roughly two or three times the depth of the unstable layer.

For streets to persist, the clouds should evaporate rapidly, and convection should not be very strong. If the clouds start to persist, rather than evaporating, they rapidly spread out beneath the inversion, giving rise to a layer of stratocumulus. If convection becomes strong, the thermals break through the inversion, and cloud growth becomes more chaotic, and the regular pattern breaks down.

Cloud streets at mid-morning, before increasing convection created larger clouds and broke up the regular pattern

BANNER CLOUDS

Occasionally, a highly distinctive plume of cloud hangs like a flag behind a mountain peak. This is known as a banner cloud or (occasionally) as a pennant cloud, and is a form of orographic cloud.

SEE ALSO
formation of
clouds (p. 86)

pressure (p. 180)

valley wind (p. 162)

Although seen on many mountains, the most famous banner clouds are those of the Matterhorn in the Alps above Zermatt on the Italian-Swiss border, and the vast plume observed stretching from Mount Everest in the Himalayas.

Banner clouds appear most frequently on isolated peaks, but there seem to be two principal effects that play a part in their formation. First, there needs to be a fairly strong wind blowing from the shadowed (and thus cooler) face of the mountain towards the side warmed by the Sun. The wind creates an increase in pressure on the windward side, and a corresponding lower-pressure region to leeward, immediately behind the peak. Second, eddies develop behind the peak, which curl both over its top and around its sides. These tend to drag air up the warmer face towards the peak of the mountain. The fact that the air is being warmed, and thus has a tendency to rise, also helps in this process.

The rising air expands (assisted by the pressure drop behind the peak), cools, and reaches its dewpoint, causing the water vapor to condense into cloud droplets. Although the cloud tends to cast a shadow on the peak, and thus reduce the warming by sunlight, an equilibrium is normally established, so that the cloud persists throughout most of the day, generally disappearing only with a major change in wind direction, or when warming ceases towards nightfall.

The Matterhorn's banner cloud, here produced by a northwesterly wind

WAVE CLOUDS

SEE ALSO

altocumulus translucidus (p. 52)

billows (p. 90)

orographic clouds (p. 87)

pileus (p. 72)

stratocumulus lenticularis (p. 36)

altocumulus lenticularis (p. 50)

cirrocumulus lenticularis (p. 66)

stratocumulus perlucidus (p. 38)

Wave clouds forming over a mountain range. The closed circulation downwind (a 'rotor') is not always present

When air is forced over a mountain barrier, it may create a series of standing waves that stretch downwind. Frequently, highly characteristic clouds form in the crests of the waves.

Wave clouds are a specific form of orographic clouds and arise when there is a stable layer (or layers) at an appropriate height. Rather than simply rising over the hills or mountains and sinking down the other side, the wave motion of the air persists, producing a long train of waves. These lee waves are present even when no clouds form and may be an unexpected hazard to aircraft.

The spacing and amplitude of lee waves depend on various factors: slower wind speeds and greater stability cause shorter wavelengths (closer spacing of the crests and troughs); and the greater the height of the mountain, the greater the amplitude. The precise nature of the waves is also affected by the general topography: both the wavelength and the amplitude may be altered by the effect of neighboring mountains, and this effect varies with wind speed. Generally, however, the waves remain almost stationary, while the wind speed, wind direction, and stability remain constant. Naturally there is a tendency for the amplitude and spacing of the waves to die out downwind, but the waves are often detectable tens (or even hundreds) of miles away.

When moist layers are involved, clouds may form in the crests of the waves. These are lenticular (lens-shaped) clouds, namely stratocumulus, altocumulus or cirrocumulus lenticularis. Such clouds are common in mountainous areas. As mentioned earlier, despite being classed as cumuliform clouds, convection is not involved, and they form only when the layers of air are stable. Although the clouds themselves appear to be stationary, the air is actually flowing through them. The clouds are continuously forming at the upwind edge, and dissipating (evaporating) on the downwind side. The clouds have many features in common with the accessory cloud known as pileus.

At times, there may be several humid layers, one above the other and it is then quite common for one lenticular cloud to appear vertically

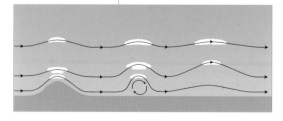

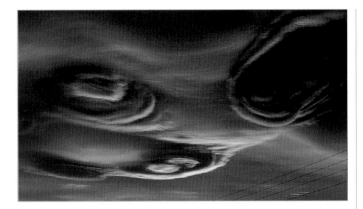

above another. When the layers are closely spaced, a spectacular stack of clouds may occur. These are known by the French term 'pile d'assiettes' ('pile of plates'). When illuminated by low sunlight, they may have a highly dramatic appearance.

Although the overall shape of wave clouds is lenticular, the clouds themselves may show all the varieties seen in normal stratocumulus, altocumulus or cirrocumulus clouds. They may be broken into individual cloudlets and thus be of the perlucidus variety, and are commonly so thin that they may be classed as translucent (translucidus). The lacunosus variety, in which holes are surrounded by cloud is rare, but not completely unknown, and lenticular clouds often consist of a series of billows (the undulatus variety). Corrugations parallel to the wind direction are also fairly common.

A pile d'assiettes over the Mourne Mountains in Ireland

FALLSTREAK HOLES

Occasionally, when a thin layer of cloud covers the sky, an isolated hole suddenly appears in the cloud. Sometimes the holes are remarkably symmetrical: appearing as almost perfect circles or ellipses. These are known as fallstreak holes.

Glaciation in high altocumulus has created this fallstreak hole and its associated dense cirrus

These holes most frequently appear in thin altocumulus, although they have also been observed in cirrocumulus. These clouds consist of supercooled water droplets—that is the cloud droplets are unfrozen, despite being at a temperature of less than 32°F. Freezing suddenly begins in part of the layer, and spreads, with the resulting ice crystals falling to lower levels. Frequently wisps of cirrus may be seen beneath a hole, but sometimes the ice crystals melt and evaporate as they reach a lower, warmer layer, leaving no sign of their presence except for the hole itself.

It is uncertain what initiates glaciation (freezing) in the cloud, one suggestion being that it is started by ice crystals falling from a yet higher, and effectively invisible, layer of cirrus. Glaciation appears to start at a single point and propagates outwards, often in a fan-shaped pattern. The exact mechanisms remain obscure, but it is thought that some of the larger droplets may fragment as they freeze, scattering tiny pieces of ice into the surroundings, which then act as freezing nuclei for other liquid droplets.

A somewhat similar phenomenon is observed when an aircraft causes a lane of thin cloud to disperse, in what is known in a dissipation trail (distrail). In this case, however, it is easier to account for the initiation of freezing through the emission of particles by the aircraft's engines.

SEE ALSO

altocumulus (p. 46)

cirrocumulus (p. 64)

distrails (p. 101)

supercooling (p. 150)

virga (p. 78)

PYROCUMULUS

Clouds will form over any significant source of heat and moisture. They are commonly observed above the cooling towers of power stations, but perhaps even more striking are the clouds that are created by fires.

Known by the rather clumsy compound formed by the Greek word for 'fire' and the Latin word for 'heap', pyrocumulus clouds vary greatly in size. They may form above quite small fires, and are common in those parts of the world where stubble-burning is still permitted. These generally take the form of small cumulus clouds at the top of the column of smoke. Because of the smoke, the base of such clouds is often darker than that of normal cumulus, or has a brownish tinge. In some cases the light smoke particles strikingly reveal the normally invisible circulation of air at the base of the cloud.

SEE ALSO

cumulonimbus
(p. 68)

cumulus congestus
(p. 24)

thunderstorms
(p. 166)

Cumulus clouds forming above wild-fires that raged in Florida in 1998

Over larger fires, such as those caused deliberately in the destruction of forests or natural wildfires, extremely large clouds may be created. These may develop as far as the cumulus congestus stage, and give rise to rain. Some reach so far into the atmosphere that they turn into cumulonimbus, not only producing heavy rainfall, but also turning into thunderstorms. Such rain has sometimes helped to extinguish the fire, but generally the cloud has drifted too far away for it to have an effect. Under extremely dry conditions, lightning from a cumulonimbus produced over a wildfire, has started new fires a considerable distance away.

JET-STREAM CLOUDS

The high-speed bands of wind that are the jet streams have an important influence on the development of weather systems, particularly the growth, motion and decay of depressions. Although often invisible, the jets are sometimes marked by a readily recognizable pattern of clouds.

The jet streams that concern us here are the strong westerly winds that lie at the top of the troposphere, close to where there are breaks in the level of the tropopause, the inversion at the base of the stratosphere. These breaks occur near major fronts, particularly near the polar front, across which there is a marked temperature contrast and on which most depressions form.

Jet-stream cirrus, showing both long streaks (radiatus) aligned with the wind, and billows (undulatus) perpendicular to it

Jet-stream clouds are generally forms of cirrus (because of the low temperatures that prevail at their height), and the clouds are often elongated along the length of the jet, thus appearing as cirrus radiatus. But jet streams are also regions of extremely high wind shear between the jet itself and the surrounding, slower moving air, so the cirrus often occurs in the form of billows (cirrus undulatus), which run at right-angles to the main direction of flow.

Jet streams are defined as having wind speeds in excess of 100 knots (about 110 mph) and may reach far higher speeds. Because of these high speeds, and despite their altitude of several miles, the movement of jet-stream clouds is usually readily detectable visually. As mentioned, jet

Bands of fast-moving jet-stream cirrus observed at sunset, and ahead of a deep depression that arrived the following day

streams have a profound effect on the development of depressions, and often lie in an 'S-shaped' curve above the main central region of a depression. The different directions in which the clouds are moving at different heights is an excellent indicator of an approaching (and receding) depression.

A jet often lies at the forward edge of the advancing wedge of warm air, approximately parallel to the warm front. At high levels the wind veers (in the northern hemisphere), relative to the lower wind. This leads to the formation of 'crossed winds': winds at high and low levels being roughly at right-angles to one another. This is a classic sign of an approaching warm front. At low levels, cumulus clouds (often cumulus humilis) may be carried along by a southerly wind ahead of the surface front, while high jet-stream cirrus (usually cirrus fibratus, cirrus uncinus, and cirrostratus fibratus) may be seen streaming away, perhaps towards the east. When the speed of the jet stream is particularly high, there may be a dramatic change in direction close to the jet-stream level.

The same sort of change in wind direction occurs with the clouds at the actual warm front: the winds carrying the middle level altocumulus and altostratus are veered relative to the nimbostratus, stratocumulus, or other low clouds. Similarly, the wind veers yet again between the middle levels and the higher jet stream.

A similar change, but in the opposite direction, occurs in the region of the cold front, where again the jet steam is roughly parallel to the front. Here the wind backs and increases with increasing height. The low clouds may be borne on a northwesterly wind, while the high cirrus races away to the northeast, carried by a southwesterly jet.

SEE ALSO

backing and veering (p. 157)

billows (p. 90)

cirrus (p. 54)

depressions (p. 180)

Polar Front (p. 176)

stratosphere (p. 188)

tropopause (p. 188)

troposphere (p. 188)

wind shear (p. 188)

CONTRAILS AND DISTRAILS

SEE ALSO

anticyclonic
weather (p. 182)

cirrus (p. 54)

fallstreak
holes (p. 96)

mamma (p. 76)

supercooling (p. 150)

warm front (p. 178)

Condensation trails (contrails) left by aircraft are a familiar sight to everyone nowadays. Although they might not seem to have much to do with the weather, they can provide useful clues to the nature of the air at height, and of the way in which the weather may develop.

Contrails are, of course, simply lines of cloud, mainly caused by the condensation of the water vapor emitted by an aircraft's engines. Water vapor itself is invisible, so there is always a clear gap behind the aircraft before the trail begins.

If you are viewing a contrail from underneath, you will notice that the trail always begins as a pair of rolls, which generally persist throughout the whole length of the contrail. These two cloud rolls do not occur, as one might imagine, because the engines are normally in pairs. In fact, they mark the presence of a pair of large, and normally invisible, vortices that are shed by the ends of the wings. The exhaust gases are sucked into the center of these two vortices, where condensation then occurs, producing the twin trails.

This contrail clearly shows the bulges that often occur on the underside of trails

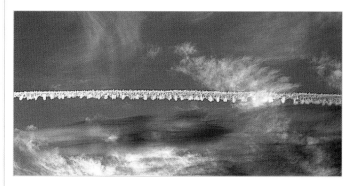

When seen from the side, contrails often display a series of downward bulges, somewhat like the supplementary cloud features known as mamma. These occur because air is being forced downwards (this is what keeps the aircraft up), but the air is also being heated by the exhaust gases, so it is buoyant and tends to rise. These conflicting forces may be likened to 'upside-down convection' similar to that found in mammatus clouds.

If the air at the flight level is dry, contrails tend to dissipate, and are sometimes extremely short. This is particularly the case if the air is subsiding, as occurs under anticyclonic conditions. If, however, the air is humid, then the trails persist and may spread out into broad bands of cloud. Generally, the water vapor freezes into tiny ice crystals, in effect creating long bands of cirrus. Persistent contrails often indicate that warm, humid air is encroaching overhead, which may mean that a depression's warm front is approaching.

Distrails

Slightly less common than contrails, but still seen fairly often, are dissipation trails (distrails). These occur when an aircraft flies at the same level (or slightly above) a thin layer of cloud, and creates a clear lane.

There are at least three ways in which this can occur. The aircraft may mix warm air from above the cloud down into the layer, causing the cloud to evaporate. Second, the heat from the aircraft's exhaust may be just sufficient to evaporate the cloud droplets. Finally, the original cloud may consist of supercooled droplets. Solid particles emitted by the engines may act as freezing nuclei and enable the supercooled droplets to freeze and fall out as ice crystals. Wisps of thin cirrus may sometimes be seen below the distrail when this has happened, so it resembles a linear fallstreak hole.

These trails are persistent, have become glaciated, and are spreading, suggesting that a warm front may be approaching (ABOVE)

An aircraft passing through this sheet of altocumulus has initiated glaciation, leaving a clear lane with a line of cirrus to its right

There is a whole range of optical phenomena that may be observed in the sky. Some are strongly colored, some exhibit pastel shades, and yet others are colorless. Some, such as ordinary rainbows, are very common, whereas others (including one form of rainbow) are so rare that any observation should be carefully recorded.

In general, optical effects may be produced by light from either the Sun or the Moon, although those formed through moonlight are often fainter and exhibit less or no color. In this initial discussion, the Sun is assumed to be the light source.

Because different optical phenomena may sometimes produce similar effects, it may need some care to distinguish between them. This is usually possible, often by a process of elimination, if sufficient details are noted. The position and angular size are frequently the most important. Angles may be estimated by the methods described on pp. 6-7. If you see some unusual optical effect, try to check the following points:

- POSITION IN THE SKY, WHETHER AROUND THE SUN OR OPPOSITE THE SUN
- ANGULAR DIAMETER (OR RADIUS), OR ANGULAR DISTANCE FROM THE SUN
- COLORS (IF ANY)
- IF DISPERSED INTO SPECTRAL COLORS, WHETHER RED OR VIOLET IS CLOSEST TO THE SUN
- IF A SPECTRUM, WHETHER THE COLORS SPREAD OUT HORIZONTALLY, VERTICALLY, IN AN ARC CENTERED ON THE SUN, OR IN SOME MORE COMPLEX MANNER
- IF A LINE OR ARC, WHETHER STRAIGHT OR A PORTION OF A CIRCLE, ELLIPSE, OR MORE COMPLEX SHAPE

Some optical effects are extremely complex and cover a large extent of sky, but as an approximate guide, they may be roughly subdivided according to their position relative to the Sun.

- **AROUND THE SUN:** CORONAE; BISHOP'S RING
- **NEAR THE SUN:** IRIDESCENCE
- **SOME DISTANCE FROM THE SUN:** CIRCUMZENITHAL ARC; CIRCUMHORIZONTAL ARC; HALOES; PARHELIA (MOCK SUNS); PARHELIC CIRCLE; SUBSUNS; SUN PILLARS
- **DIRECTLY OPPOSITE THE SUN (AT THE ANTISOLAR POINT):** GLORIES; HEILIGENSCHEIN
- **CENTERED ON THE ANTISOLAR POINT:** RAINBOWS; FOGBOWS; DEWBOWS

The 22° halo in thin cirrostratus (FAR LEFT)
A double rainbow visible in the spray from a fountain (BELOW)
A mock moon (paraselene) is not particularly rare, but often goes unnoticed (BOTTOM)

SEE ALSO

color of the sky (p. 128)

cloud colors (p. 134)

sunset and sunrise effects (p. 130)

crepuscular rays (p. 136)

shadows (p. 138)

CORONA

A corona is an optical phenomenon immediately surrounding the Sun or Moon, as seen through thin cloud. Because of the Sun's brightness, casual observers are more likely to see it around the Moon, but hiding the Sun will often reveal a corona. It occurs most frequently in thin altocumulus and altostratus but occasionally in other cloud types.

The aureole, or innermost region of a corona, has a reddish edge

Coronae consist of an inner aureole, outside which there may be one or more sets of colored rings. The aureole and rings arise through diffraction of the light by the cloud particles. In diffraction, light deviates from its original path when it encounters small particles, being bent round them, rather than passing through the particles. Any colors are produced by interference effects, because the light reaching the observer has taken various paths of different overall lengths. The colors and their purity are strongly dependent on the size and uniformity of the cloud particles. Uniform particle sizes produce the purest colors.

The inner aureole of a corona is a bluish-white disk of light immediately around the Sun or Moon, with a brownish-red outer edge. On relatively rare occasions, a blue or violet tint may be seen in the center. When the cloud particles have a wide range of sizes, the aureole may be the only portion of the corona that is visible.

When the cloud particles are uniform in size, one or more outer rings of spectral colors may become visible, with violet on the inside, and red on the outside. The radii of these rings are strongly dependent on the particle sizes: the smaller the particles, the larger the resulting radii.

Because of variations in the distribution of particles within a cloud, coronae may be irregular in outline, and tend to alter in shape and coloration as clouds pass across the Sun or Moon. If no changes are apparent, it is safe to assume that the cloud particles are extremely uniform in size, and evenly distributed.

Coronae are by no means confined to cloud droplets. They may occur whenever there is a cloud of evenly sized particles, as with mist or fog—or even when wearing misted-up eyeglasses. The vast clouds of pollen shed by certain trees and shrubs in spring may also produce them, and they are quite commonly observed with the pollen from birch and pine forests, for example.

On very rare occasions, when the Sun is high in the sky, it may be surrounded by a pale disk with a faint reddish-brown outer border, which has inner and outer radii of approximately 10° and 20°, respectively. Although this effect resembles the central aureole of a corona, no clouds are visible. It is, basically a diffraction corona, but caused by tiny particles of dust or sulphur dioxide, which have been lifted into the stratosphere by violent volcanic eruptions. The ring was first reported from Hawaii after the famous eruption of Krakatau in 1883, by the Reverend Bishop, after whom it is now named. It has been seen on a few occasions since then, and some observers have reported a bluish tinge to the disk, but this seems to be uncommon.

SEE ALSO

blue Sun (p. 129)

purple light (p. 133)

Spectral colors appear in the outer region of a corona

GLORY

SEE ALSO
Brocken Spectre
(p. 141)

fogbow (p. 114)

The optical phenomenon known as a glory consists of a series of colored rings that appear around the antisolar point, when the sunlight is falling on a cloud or a bank of mist or fog. It is now most commonly seen from an aircraft, but is also often noticed by people standing on a mountain ridge above a layer of cloud or fog. In such cases, individuals see a glory around the shadow of their own head, but not around those of any companions. Similarly, if seen from an aircraft close to a bank of cloud, the position of the glory relative to the shadow of the plane will depend on where the observer is sitting. At greater distances, the plane's shadow becomes indistinct and the glory appears on its own.

The colors in glories are very similar to those seen in coronae, with violet in the inside, and red on the outside. Again, multiple rings may occur and, as with other diffraction phenomena, the smaller the cloud particles, the larger the radii of the rings. Although the general principles involved in the creation of a glory are understood, a rigorous mathematical explanation has yet to be discovered.

A colored glory should not be confused with a bright, but colorless heiligenschein, which also appears around the shadow of the observer's head.

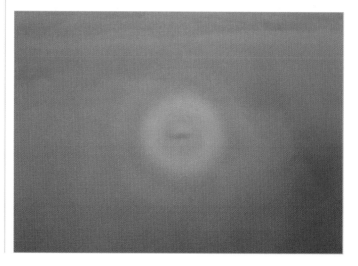

A glory surrounding the shadow of the observer's aircraft

HEILIGENSCHEIN

The heiligenschein is a bright halo of light that appears around the shadow of the observer's head or around that of the camera (i.e., at the antisolar point). The name means 'holy light' in German and, as with a glory, the observer sees only the light surrounding his or her own head, not around those of any companions.

SEE ALSO
dew (p. 146)

dewbow (p. 115)

guttation drops (p. 147)

A dewdrop heiligen-schein appearing around the shadow of the photographer's head

There are, in fact, two mechanisms that produce the heiligenschein effect. The most common occurs when sunlight falls on dew-covered grass, and indeed grass is more effective in producing the effect than other types of vegetation. Many species of grass are covered in short hairs, and these support the dewdrops slightly above the surface of the leaf. The result is that the dewdrops and leaf together act as an almost perfect reflector (rather like the 'cat's eyes' used on roads), returning light back towards its source, and thus producing the bright halo.

The other effect does not require the presence of dew, but may be seen on a patch of grass, a field of grain, or any other surface that consists of innumerable individual elements. When looking in precisely the same direction as the incident light, each individual blade or leaf is fully illuminated, but its shadow is obscured by the object itself. So all the individual objects appear bright. Farther away from the antisolar point the shadows begin to appear, so the overall brightness decreases away from the center.

If observers are sufficiently far away, any shadows that they cast become relatively small, and the bright spot becomes all the more conspicuous. This bright spot is often visible from an aircraft—it is known as the 'hot spot' to aerial photographers—and may be observed apparently gliding across the fields and woods below.

IRIDESCENCE

Iridescence is one of the most common, yet is probably also the most overlooked of all optical phenomena. It most frequently appears as bands of color around the edges of thin clouds, particularly altocumulus, altostratus, and cirrocumulus. It also sometimes occurs in thinner patches within a layer of cloud, and is also seen around the edges of clouds as they pass across the Sun. The colors are often delicate pastel shades.

Beautiful coloration is visible in this iridescence in very thin cirrus cloud

Like coronae, iridescence is produced by diffraction. Unlike them, it may occur at a range of distances from the Sun, although it is often strongest at an angular distance of 30–35°. Because of the way in which the colors are produced, iridescence always appears on clouds in the same general direction as the Sun, and this is why it goes unnoticed: it is often lost in the general glare, or overwhelmed by the brightness of neighboring clouds. Shielding your eyes from direct sunlight will help, as will the other methods described elsewhere under 'Observing the Sky' (pp. 6-7). The iridescence produced by moonlight is often easier to see, but seems to be largely ignored.

A band of a particular color shows where all the cloud particles (which may be water droplets or ice particles) are approximately the same size. The purity of color depends on how uniform the particles are in diameter. When a range of sizes is present in the one region of the cloud, light of slightly different colors is mixed together, resulting in a paler, pastel shade.

The most striking examples of iridescence are seen in nacreous clouds, but these are relatively rare. Similarly strong iridescence is also visible occasionally in persistent trails that have been created by research rockets in the lower stratosphere.

SEE ALSO

corona (p. 104)

nacreous clouds
(p. 80)

Iridescence in thin trails of ice crystals at the top of a large cumulonimbus

Pastel-colored iridescence at the edge of a layer of altostratus

RAINBOWS

Although rainbows are very common, and nearly everyone has seen one, there are a number of features, including some that are quite rare, that often go unnoticed. Probably the most favorable conditions for the formation of rainbows are showery conditions: what the forecasters sometimes describe as 'sunshine and showers'. Such showers tend to be less frequent early in the day, before convection has become fully established, so rainbows tend to be seen later in the day.

This double rainbow clearly shows the primary (inner) and secondary bows, separated by Alexander's dark band, together with supernumerary bows within the primary bow

Rainbows are produced when sunlight falls on raindrops and is then both reflected in the direction of the observer, and also dispersed into the spectrum. Because the light is reflected inside the raindrops, rainbows appear on the opposite side of the sky to the Sun. They are always portions of circles, centered on the point directly opposite the Sun, known as the antisolar point. The Sun, the observer's head (or the camera) and the antisolar point always lie on a straight line. The amount of the circular arc that is visible depends on several factors. Under perfect conditions it appears to stretch right across the sky, and 'touch' the ground at each end. The height of the center of the arc depends on the altitude of the Sun: the higher the Sun, the lower the rainbow. When the Sun is on the horizon, the rainbow is a perfect semi-circle. From an aircraft or, very occasionally, from a high vantage point such as a mountain, a complete circle may be seen.

Very often, only part of a rainbow is visible, usually one end or the other. This happens when just some of the raindrops are illuminated by sunlight (the others being perhaps in shadow), or when rain is falling in only part of the sky, and there are no drops to form the rest of the bow. When just a small part of a bow is visible, it is sometimes mistaken for another optical phenomenon, such as part of a halo or a parhelion (mock sun). These phenomena occur in the part of the sky towards the Sun, however, not away from it. In any case the type of cloud is an immediate indicator: precipitating cumulonimbus or cumulus congestus produce rainbows, whereas ice crystals, and thus cirriform clouds, are required to create halo effects.

The most commonly seen form of rainbow, known as the primary bow, has a radius of approximately 42°, and the spectral colors always appear with red on the outside and violet on the inner edge. This type of rainbow appears when the light is reflected just once by the rear of each raindrop before it is returned towards the observer's eyes. Quite

This rainbow was photographed just before sunset, when only the red and orange wavelengths are able to penetrate through the atmosphere

frequently, however, some of the light is reflected twice within the raindrops before being sent back towards the observer. This second set of raindrops produce a secondary bow, which has a larger radius of approximately 51°—and thus appears outside the primary bow—in which the sequence of colors is reversed, so red appears on the inside and violet on the outside edge. The secondary bow is normally fainter than the primary bow.

When large portions of both primary and secondary bows are visible, it is often obvious that the region between the bows is much darker than the surrounding sky. This region is known as Alexander's dark band, and occurs because the raindrops in that area are sending the light in a different direction, and not towards the observer.

Under good conditions it is also often possible to see one or more bows lying inside the primary bow. Such bows are known as supernumerary or interference bows, and occur when light takes paths of slightly different lengths through the raindrops, giving rise to interference colors like those seen in a film of oil on water. These bows are generally pale and do not show such strongly saturated colors as those in the primary and secondary bows. To most observers they commonly appear pinkish-violet and greenish. On very rare occasions, however, the colors may be almost as strong as those in the main bows. Supernumerary bows may be separated from the inner (violet) edge of the primary bow, or from one another, by a narrow dark band or bands.

Because the primary and secondary bows are centered on the antisolar point, and that is determined by the position of the observer, everyone sees their own personal set of bows. (A similar effect occurs with glories and the heiligenschein.) As the observer moves, so does the bow, hence the impossibility of reaching the 'pot of gold at the end of the rainbow'. More striking, however, is the effect that can occur if there is a reflecting surface, such as a lake or other body of fairly calm water, between the Sun and the observer. Sunlight reflected from the lake may strike the raindrops in a completely different direction from that falling directly on them without reflection. The reflected light produces another set of rainbows, lying in the same general direction as the normal bows, and with precisely the same radii, but which are centered on a point higher in the sky. Under good conditions, therefore, these reflected-light bows consist of much larger circular arcs, but because the radii are identical, the pairs of primary and secondary arcs appear to intersect. When only parts of the bows are visible, it may seem as if the primary or secondary arcs (or both) split into two. The strength of such reflected-light bows depends on the angle at which sunlight strikes the reflecting surface. For a low Sun, the amount reflected may be very high (as much as 95 percent), and the bows may be nearly as bright as the rainbows produced by direct illumination.

The strength of color seen in rainbows is affected by the size of the raindrops. When the droplets are very large, the colors are strong, and red is prominent. With smaller droplets the colors are less saturated and the outermost color of the primary bow tends towards an orange shade, rather than a true red. The presence or absence of particular colors in supernumerary bows, and of the dark bands between them, is closely linked to droplet size, but in far too complex a manner to be discussed here.

The colors visible are also affected by the altitude of the Sun. Towards sunset, sunlight becomes greatly reddened, with the shorter wavelengths having been scattered away. Yellow-orange-red and even all-red rainbows have been observed at such times.

A low double bow, photographed about noon, when the Sun was high in the sky

FOGBOWS

With the very finest droplets, such as those found in mist or fog, the light is no longer reflected and refracted within the drops, but is diffracted by them instead. Nearly all the color disappears and a white fogbow is seen. This has the same radius as the usual primary bow (approximately 42°). Occasionally the inner and outer edges show faint bluish and reddish tinges, respectively, as with a normal rainbow. Such fogbows are sometimes accompanied by a glory. When seen from inside a cloud, or from an elevated position, a complete circle may appear. As seen from an aircraft, cloud droplets may produce a similar bow, sometimes called a cloudbow, which appears to lie on the surface of the clouds. Depending on the exact conditions this may appear as a closed ellipse, or an arc (either a parabola or hyperbola) where the open arms stretch away from the observer.

A typical colorless fogbow

MOONBOWS

The Moon also produces rainbows, but because its light is much fainter than that from the Sun, the colors are extremely pale and a lunar rainbow often appears white to an observer. Because moonlight simply consists of reflected sunlight, however, all the colors are actually present. Long-exposure photographs reveal precisely the same colors in both the rainbow and the surrounding landscape and therefore cannot be distinguished from photographs taken in daylight, except perhaps by the blurring of objects in motion.

Bows identical to rainbows occur with water droplets from waterfalls, fountains, garden hoses, and even in the spray from vehicles. Apart from the primary and secondary bows, higher-order bows (with a greater number of reflections within each droplet) may be observed in laboratory experiments. The faint tertiary (third-order) bow actually appears on the same side of the sky as the Sun. There have been reports that it has been seen, but these have never been reliably confirmed.

DEWBOWS

The droplets of dew may also act in the same way as raindrops and produce a colored bow—a dewbow. This is most frequently seen on grass-covered areas in autumn, when hundreds of tiny spiders' webs, spun between the blades of grass, hold dewdrops suspended above the ground. As with a cloudbow, because it is seen on a more-or-less horizontal surface, a dewbow does not appear as a circular arc, but as an ellipse or as a parabolic or hyperbolic arc. Generally just one or two portions of the arc are visible, lying to the left and right of the antisolar point. A similar bow has been observed on the surface of a pond, where it is caused by droplets of water resting on the surface film.

SEE ALSO

blue of the sky
(p. 128)

cirriform clouds
(p. 15)

cumulonimbus
(p. 68)

cumulus congestus
(p. 24)

dew (p. 146)

haloes (p. 116)

showers (p. 164)

sunrise/sunset
colors (p. 132)

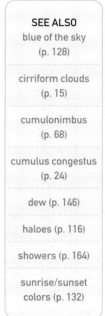

An unusually strong dewbow, created by droplets hanging from spiders' webs woven between the blades of grass

HALO PHENOMENA

A complex
halo display,
photographed from
the South Pole

The presence of ice crystals in the atmosphere produces a whole range of optical effects. Because the crystals may refract the light, dispersing it into a spectrum, or simply reflect it, some of the effects are strongly colored, while others are colorless (i.e., white). The positions of these circles, arcs, and spots of light are strictly governed by the shapes that ice crystals may assume. Nevertheless, there are so many forms that not all the rare types are fully understood.

Both 22° and 46° haloes are accompanied by a parhelic circle, two parhelia (on either side of the Sun), the upper tangent arc (touching the 22° halo), and a suncave Parry arc

HALOES

Haloes are circular arcs around the Sun (or Moon). They are actually extremely common, being visible on perhaps one day in three in Britain and western Europe. This is because (among other occasions) they occur whenever there is a thin veil of cirrostratus, and this often precedes a warm front. But under such conditions, the sky remains bright, and the majority of people do not even realize that clouds are beginning to cover the sky.

In temperate regions there tend to be more reports of 'rings around the Sun' in summer, when sun-bathers feel that the warmth of the Sun has

A 22° halo in slightly fibrous cirrostratus

started to decline, hold up their hands to block the sunlight, and see a halo for the first time in their lives. Yet haloes occur throughout the year and are particularly common in the polar regions, and in the winter in continental regions, where the cold conditions often produce diamond dust, tiny ice crystals that float in the air and create brilliant haloes.

The most frequent halo is a circle around the Sun or Moon, with a radius of 22° (a handspan at arm's length). It is produced when light passes through two (non-adjacent) faces of a hexagonal ice crystal, with an included angle of 60°. These crystals normally have completely random orientations so that the light is spread in every direction and thus appears as a circle at a fixed distance from the Sun. Because of variations in the amount and nature of the cloud that is present, the circle is often incomplete, or varies in strength.

When faint, this ring appears white, but when stronger, the inner edge is red, surrounded by a yellow tint. On rare occasions a blue-violet coloration has been reported on the outside. This sequence of colors is exactly the opposite of that seen in a primary rainbow, but is hardly necessary as a means of identification, because a halo lies on the same side of the sky as the Sun, whereas a rainbow always appears around the antisolar point. The sky inside the halo is darker than that outside. The reason, as with Alexander's dark band in a rainbow, is that light from that portion of the sky is diverted away from the observer.

A second halo, with a radius of 46°, may sometimes be seen. This occupies such a large area of the sky that it is rare to see a complete circle. It is much fainter than the 22-degree halo, which is undoubtedly another reason why it is less often reported. It is produced when light passes through the end face and one side of a crystal where the included angle is 90°.

PARHELIA

A bright spot of light that appears on one side of the Sun is known as a parhelion, more commonly called a mock sun, false sun, or sun dog. Such parhelia are the most common halo phenomenon after the 22-degree halo. Frequently only one is visible, but it is also quite common to see one on each side, and at the same altitude as the Sun. Their exact position relative to the 22-degree halo depends on the Sun's altitude. When the Sun is low, they lie on the halo. The higher the Sun, the farther the parhelia lie outside the 22-degree halo, reaching about 5–6° for a solar elevation of 45°. Parhelia disappear, however, when the Sun's altitude exceeds approximately 60°.

Parhelia are produced when light is refracted through normal hexagonal ice crystals in the form of flat plates, where the large flat surfaces are lying horizontally. Plates of this sort tend to adopt this position as they fall through the air. Because of the concentration of one particular orientation, the parhelia may be brilliantly colored and clearly show all the normal spectral colors. This form is often observed in isolation—i.e., without the 22-degree halo—in small patches of cirrus or, occasionally, in ice-crystal virga beneath other types of cloud.

A brilliant parhelion in an isolated patch of cirrus

The size of parhelia is often at least as large as the Sun, or even larger, so they are quite conspicuous. In addition, when close to the 22-degree halo, they tend to appear reasonably circular, but for higher solar elevations, they tend towards a diamond shape. The brightest parhelia have white 'tails' that stretch away from the Sun, but still at the same altitude. These tails arise because light strikes ice crystals at different distances from the Sun at slightly different angles. The fans of refracted light overlap, so the colors are lost with the exception of the parhelion's red inner edge. The tails lie along the same line as another halo effect, the parhelic circle, described shortly.

The Moon may produce a similar effect. Such a mock moon is known as a paraselene (pl. paraselenae).

CIRCUMZENITHAL ARC

The purest spectral colors shown by any halo effects occur with the two arcs known as the circumzenithal and circumhorizontal arcs. The first of these occurs only at low Sun elevations of approximately 5–30°. It is a bright arc, a portion of a circle that is centered on the zenith—the point directly above the observer's head. This arc is produced by the same type of crystals that create the 46-degree halo, and its position is closely related to the location of that halo, even though the latter may not be visible at the same time. The exact location, length, and brightness all vary with the Sun's actual elevation.

A strongly colored circumzenithal arc in thickening cirrus ahead of a warm front

At low elevations, the arc is faint, and lies between the 46-degree halo and the zenith. With increasing elevation, the arc lengthens and strengthens, until at a solar elevation of 15° it is at its longest, forming slightly less than one-third of a full circle—actually 108°—and is fairly close to the position of the 46-degree halo, which it touches at an elevation of 22°. As the elevation continues to increase, the arc moves back towards the zenith and shortens again. It is always convex towards the Sun, and strongest at the point closest to the Sun.

As with many other halo effects, the circumzenithal arc may appear on its own in an isolated patch of cloud. It is actually more common than the 46-degree halo.

CIRCUMHORIZONTAL ARC

The brilliantly colored circumhorizontal arc runs parallel to the horizon. It is closely related to the circumzenithal arc and is produced by the same crystals, but is visible only when the Sun's elevation exceeds 58°. (The limits are approximately 58–80°.) This is because it lies below the lowest point of the 46-degree halo. It cannot, therefore, be seen at all from high latitudes, and is visible only in high summer from southern Britain, middle and southern Europe, and the northern United States. It is more readily seen at low latitudes, and from most inhabited areas of the southern hemisphere, except the southernmost tip of South America.

A brilliant circumhorizontal arc observed in thick cirrus

Again, its precise relative position varies with solar elevation, touching the 46-degree halo when the Sun's altitude is 68°. Its distance from the halo increases towards both lower and higher elevations. If anything, its brilliant colors are even more spectacular than those of the circumzenithal arc and, like that arc, it may extend for nearly one-third of the way around the horizon.

PARHELIC CIRCLE

A white arc that runs around the sky, parallel to the horizon, and at the same altitude as the Sun is known as the parhelic circle. It is a reflection effect. As the flat, plate-like crystals that produce parhelia fall through the air, their large flat sides are approximately parallel to the ground, and the shorter sides are vertical. Light reflected from these short sides is projected all round the sky to produce the parhelic circle. However, it should be said that a full circle is rarely seen outside the polar regions or continental regions in winter, although short segments are moderately common.

A bright parhelion and portion of the parhelic circle

SUN PILLAR

Another reflection effect, the sun pillar, appears as a vertical column of light that passes through the Sun. This may be produced either by flat plate crystals or by pencil-shaped ones. There are subtle differences in the overall shape of the two types of pillar, but these are not often apparent when viewing from the ground, when the lower portion of the pillar is invisible. If, from an aircraft, the pillar has the appearance of an extremely elongated 'figure of eight' filled with light, it is being produced by flat plates. This shape of each half, above and below the Sun, is essentially the same as that displayed by a glitter path. If the sun pillar appears straight-sided, then pencil crystals are responsible.

A sun pillar sometimes appears at the same time as a portion of the parhelic circle to produce a cross in the sky, centered on the Sun. When the Sun is low, or even below the horizon, its light will be red, so the sun pillar appears a similar color. Care must be taken, however, not to confuse such a sun pillar with a bright crepuscular ray, which just happens to be vertical. Sometimes such a shaft of light, projected onto the base of a layer of clouds may appear remarkably like a sun pillar.

A sun pillar produced by ice crystals in cirrus undulatus

As with most halo phenomena, light from the Moon may give rise to the same effect. In cold climates, particularly when diamond dust is present in the air, it is quite common to see light pillars above streetlights and other sources of artificial illumination. Such light pillars have sometimes been mistaken for auroral rays.

SUBSUN

From an aircraft or similar high position, a brilliant spot of light is sometimes visible at the same distance below the horizon as the Sun is above. This is a very specific version of a sun pillar, known as a subsun. It occurs only when the faces of the crystals are perfectly horizontal, and thus act as tiny mirrors. The sun pillars just described are produced by crystals with faces slightly inclined to the horizontal—hence the light is spread into a column above and below the Sun. The closer the orientations become to perfectly horizontal, the more the column shrinks vertically, until it eventually becomes a bright spot located below the Sun.

A typical subsun, as observed from an aircraft

OTHER ARCS AND POINTS OF LIGHT

The effects described are the most common halo phenomena. There are many others, some of which may be briefly mentioned:

- **PARRY ARCS:** BRIGHT ARCS ABOVE AND BELOW THE 22-DEGREE HALO; VERY VARIABLE IN SHAPE, DEPENDING ON SOLAR ELEVATION. DIFFICULT TO DISTINGUISH FROM THE UPPER AND LOWER TANGENTIAL ARCS AT CERTAIN ELEVATIONS

- **ANTHELION:** A BRIGHT SPOT AT THE SAME ALTITUDE AS THE SUN, BUT ON THE OPPOSITE SIDE OF THE SKY

- **HALOES WITH UNUSUAL RADII:** 9°, 18°, 20°, AND 35° HAVE ALL BEEN REPORTED

- **ARCS OF LOWITZ:** ARCS EXTENDING FROM THE 22-DEGREE HALO, BELOW THE PARHELIC CIRCLE TO THE POSITIONS OF THE PARHELIA, PLUS ADDITIONAL ARCS ABOVE THE HALO. ALL VARY GREATLY WITH SOLAR ELEVATION

- **SUPRALATERAL AND INFRALATERAL ARCS:** ARCS ABOVE THE 46-DEGREE HALO AND BELOW IT TO EACH SIDE. AGAIN, VERY VARIABLE WITH SOLAR ELEVATION

- **SUBPARHELIA:** SEEN FROM AN AIRCRAFT, PARHELIA-LIKE SPOTS OF LIGHT 22° DISTANT ON EACH SIDE OF THE SUBSUN

- **120-DEGREE PARHELIA:** SPOTS OF LIGHT ON THE PARHELIC CIRCLE, 120° FROM THE SUN

- **ANTHELIC ARCS:** WHITE ARCS THAT CROSS AT THE ANTHELIC POINT

- **ANTHELIC PILLAR:** A COLORLESS PILLAR THROUGH THE ANTHELIC POINT

SEE ALSO

aurora (p. 84)

cirrostratus (p. 60)

cirrus (p. 54)

crepuscular rays (p. 136)

diamond dust (p. 145)

virga (p. 78)

Some of these effects, and other, even rarer ones, are still poorly understood. Good-quality photographs are of considerable scientific value. If you photograph any such phenomenon, please ensure that you note down the focal length of the lens (or setting of a zoom lens), so that angular distances may be calculated accurately from the photograph.

Upper tangential arc to the 22° halo

MIRAGES

Mirages occur when there are steep temperature gradients in the atmosphere. Air at different temperatures has different densities, and variations in density cause the amount by which light is refracted (deviated from its original path) to vary accordingly. Some deviation always occurs in the atmosphere, but in mirages it causes a noticeable distortion of images.

The effect of changes in density is to cause rays of light to deviate from a straight line and curve in the direction of the denser air. This effect occurs in the atmosphere even under normal conditions, when the air density decreases with height. At sunrise and sunset, for example, rays of light from the Sun (or Moon) are refracted by the atmosphere and curve downwards towards the surface. As a result the Sun and Moon are actually visible when they are below the geometrical horizon.

There are two main forms of mirage, classed according to whether the image of a distant object (or part of it) appears lower or higher than would normally be expected:

 INFERIOR MIRAGE: OBJECTS APPEARS LOWER

SUPERIOR MIRAGE: OBJECTS APPEAR HIGHER

Everyone will have seen the type of mirage where pools of water appear to be lying on a hot road. This is a form of inferior mirage. The 'water' is actually an image of the sky, which thus appears lower than expected.

An inferior mirage over a hot, dry-lake bed: distorted images of vehicles, Edwards Air Force Base, California

An inferior mirage caused by extremely low temperatures in Alaska. The distant mountains are unaffected, but an inverted image of the nearer hills may be seen in the center of the picture

The lowermost layer of air is strongly heated, and thus becomes less dense than the overlying air. Rays of light from the sky are bent upwards, giving the appearance of having been reflected from the hot surface.

This type of mirage actually produces an inverted image of distant objects, but this may be so distorted as to be unrecognizable. In many cases, however, a distinct, reflected image of distant objects, appears below the image of the objects themselves. This 'reflected' image is often compressed vertically.

In a superior mirage, the opposite conditions occur: there is a temperature inversion with a warmer, less dense layer of air above a colder, denser one. Rays of light are deflected downwards towards the observer and may produce an inverted image of a distant object, apparently floating in the air above the horizon. Such conditions are less frequent than those found in inferior mirages, but are frequently seen over the sea, which is sometimes much colder than the air immediately above it. This may occur after a very hot day, for example, or in spring, before the water has started to warm up after the winter. Similar conditions may occur above an expanse of ice.

Superior mirages, in particular, may produce multiple images, both upright and inverted. The atmospheric layers may be very shallow, and significant changes may occur if you bend down to get a different viewpoint. One striking effect is known as the Fata Morgana (sometimes called 'castles in the air'), when distant objects appear extremely elongated, giving the impression of tall buildings or towns in the distance. What seems to be cliffs or a high wall along the horizon is actually the distorted image of the relatively flat sea or ice floes.

SEE ALSO

inversion (p. 187)

sunset and sunrise effects (p. 130)

Quite apart from the optical phenomena that have been discussed, there are a number of effects of light and color that are visible in the sky. Even in the absence of any clouds, the sky background itself may exhibit different colors. How do these arise? We can start by understanding why the sky is blue. The color of the sky (and many other meteorological phenomena) arises from a mechanism called scattering. Molecules of gas, water-vapor droplets, dust, smoke, and any other small particles deflect light that is falling on them. The important point is that not all wavelengths (i.e., colors) are scattered equally. Those wavelengths that are scattered are strongly dependent on the size of the individual particles. The oxygen and nitrogen molecules that form the bulk of the air scatter violet and blue light, but have little effect on longer wavelengths. The scattering occurs more-or-less in all directions, and it is this scattered blue light that accounts for the color of the sky. (Human eyes are relatively insensitive to the violet light that is also scattered.)

When sunlight has to pass through a considerable part of the atmosphere, as at sunrise and sunset, practically all of the blue light is removed. Only the longer wavelengths reach the observer. This is why the rising or setting Sun is yellow or red. It is also why the Moon appears red during a lunar eclipse, because only red light penetrates right through the Earth's atmosphere and is available to illuminate the Moon.

Larger particles scatter other wavelengths of light. In particular, water vapor scatters longer wavelengths and 'dilutes' the pure blue. This is why the sky appears deep blue when seen from the top of high

When the air is humid, the sky is pale blue, and distant objects take on pale tints, giving rise to aerial perspective

The dark haze layer that has built up during the day is clearly visible in this sunset photograph

mountains, where the air is cold and dry, and a much paler shade of blue when seen from lower levels, where the air is more humid. Even at low altitudes, visibility is much greater in dry air than when there is a lot of water vapor present.

Dry haze particles also scatter longer wavelengths, as well as absorbing some of the light. When the Sun is high, haze may give a pearly quality to the light, but when the Sun is low on the horizon, the haze often appears as a distinct dark layer, with a brownish tinge. If the particles are of a very limited size range, they may selectively remove certain wavelengths. A blue Moon or blue Sun does exist. Smoke from forest fires or fine dust particles—particularly those arising from the soil known as loess, found over wide areas of China—often remove all wavelengths except blue or green, so the Sun and Moon may appear pale blue or green.

There is no water vapor in the cold, dry air at the South Pole, so the sky appears a deep blue, even close to the Sun (FAR LEFT)

There are a number of different phenomena that are visible at sunrise or sunset and which affect the appearance of both the Sun itself and of the sky around it.

When the Sun is low on the horizon, its light is passing through the densest part of the atmosphere. Not only does the air scatter all the shorter wavelengths aside, so that the Sun appears yellow or red, but it also causes distortions of the image. When the air is free from haze, refraction causes the Sun's disk to be flattened vertically. (The rays of light are also bent by the atmosphere, raising the Sun's image. When it appears to be touching the horizon, it is, in fact, geometrically below it.)

Layers of differing density often distort the image of the setting Sun

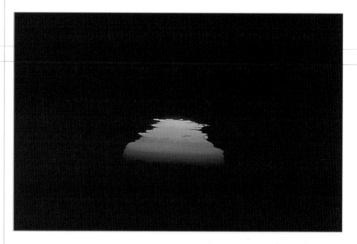

Frequently, however, the Sun's actual image is not a smooth ellipse, but has a wavy outline. This occurs when there is a series of layers of different density (i.e., of different temperature). The variations in density cause differing amounts of refraction, leading to different widths of the disk appearing at a particular altitude. This appearance is sometimes known as a 'laminated' Sun, and is related to the effects caused by mirage conditions. Sometimes, especially when there is a heavy haze layer, the final segment of Sun that is visible is actually some way above the true horizon. Naturally, these effects are also visible at sunrise, although a dense haze layer is nearly always absent in the early morning, because it tends to build up throughout the day.

The green flash

SEE ALSO

blue sky (p. 128)

mirage (p. 126)

purple light (p. 133)

When the sky is very clear, and the setting Sun may be watched as it sinks below the distant horizon, on very rare occasions the last, tiny portion of the Sun that is visible appears as a flare of brilliantly green light. This is the famous green flash. (On even rarer occasions it may appear as a blue-violet flash, rather than green.) Rather than a brief flash of light, the whole of the last segment of the Sun may sometimes appear a beautiful emerald green. This effect is sometimes known as the green segment.

Over the years there has been a great deal of discussion about the reality and possible causes of this phenomenon. There are strong arguments that suggest that the atmosphere cannot disperse light sufficiently for green (or blue) to become visible to the eye, and that the effect arises because, through staring at the Sun, the eye has become insensitive to red light, causing a shift in apparent colors, with yellow being perceived as green. There are also cogent arguments that point out that the effect is visible at sunrise, as well as sunset, when the eye has yet to see the Sun; and that both the green and blue flashes have been photographed. As a counter-argument to the last point, it is known that the color rendering of films may be extremely sensitive to the precise exposure and the exact type of film. (This problem also affects other atmospheric phenomena, in particular, photographs of the purple light.)

Whatever the true facts of the case, and even if they are optical illusions, both the green flash and the green segment are so distinctive that they leave a lasting impression on observers. A sea horizon seems to be favored, but the effect is by no means unknown over distant hills (or even buildings).

A green flash observed as the Sun sank beneath the distant horizon

The semicircular area above the position at which the Sun has set, or is about to rise, is known as the twilight arch. At times it may display extremely striking colors.

SUNSET COLORS

After the Sun has completely disappeared below the horizon, the area of sky immediately above the point where it was last visible is pale yellow, with a bluish-white segment above it. On either side, the sky is orange on the horizon, shading upwards into a yellowish tint. Slightly later, the twilight arch becomes orange at the horizon, grading into yellow, and then a salmon-pink higher in the sky. Sometimes a greenish tinge is visible on either side, where the yellow area grades into the blue of the sky. As time passes, the salmon-pink area slowly sinks towards the horizon, gradually flattening out, while the sky above slowly alters from a bluish-grey to purple-blue, and eventually to a deep blue shade. Along the horizon, the last hints of light are often yellowish-green, shading into the deep blue overhead.

Naturally, a similar, but reversed, sequence of colors is visible at sunrise. Indeed twilight colors are normally much stronger at dawn, when the sky is generally clear of haze. This is particularly the case when the sky may be

A fairly typical twilight arch above the position of the setting Sun

seen above an uninterrupted sea horizon. When there is a clear sky above the Sun, and mountains on the other side of the sky, there may be a spectacular display of the alpine glow and also of the shadow of the Earth.

THE PURPLE LIGHT

On very rare occasions, the top of the twilight arch may turn a spectacular, vibrant purple, which is far more intense than the normal purple tinge visible at twilight. This effect is normally quite short-lived, and only on infrequent occasions does it last more than a few minutes. The purple light is extremely striking visually, but is one of the few atmospheric effects that are exceptionally hard to photograph. This is because of the color response of the various layers that make up color films. (A similar problem is encountered in reproducing the color of certain flowers.) Unfortunately the purple light is seen so infrequently that it is difficult to experiment with different films to find the most suitable.

SEE ALSO

alpine glow (p. 138)

blue sky (p. 128)

shadow of the Earth (p. 139)

The purple light is normally seen only after there has been an energetic volcanic eruption, which has ejected material into the stratosphere. This may be fine volcanic dust or sulphur dioxide particles. These particles are large enough to scatter significant amounts of red light from the Sun. The distinct purple shade is actually produced by a mixture of this red light with the normal blue light that is scattered by the molecules of oxygen and nitrogen in the air. (The scattered blue light is present all the time. Long-exposure photographs taken at night show the sky as the same blue that we are familiar with during the daytime.) Although material may remain suspended in the stratosphere for many months, the exact conditions required to produce the purple light generally occur over a period of just a few days, unless the volcanic eruption has been extremely violent or long-lived.

The photographic emulsion has failed to capture the vibrant purple color visible to the eye, only the red tints having been recorded

CLOUD COLORS

There is some truth in the old rhyme of 'Red sky at night, sailors' delight, red sky at morning, sailors' warning'. Most weather systems move from west to east, so clear sky to the west allows the setting Sun to illuminate clouds overhead with red light. The clouds are moving east, and clear skies will move in overnight. Early in the morning, however, if it is clear in the east, and cloudy in the west, the weather may well deteriorate during the day.

Remnants of contrails brilliantly illuminated by the setting Sun, heralding the fine day that followed

Clouds generally consist of innumerable tiny water droplets, which are so small that they scatter light of all wavelengths and do not allow light to penetrate very far into the body of the cloud. They thus appear brilliantly white when fully illuminated by the Sun. When clouds start to decay, however, the smallest droplets evaporate first. The clouds become more transparent, allowing light to penetrate more deeply, and more to be absorbed by the large droplets. Older clouds and also the edges of clouds that are evaporating therefore often appear darker even when they are in full sunlight. Clouds that consist of ice crystals (snowflakes) not only appear softer in appearance than water-droplet clouds, but also tend to be slightly darker for the same reason.

Frequently some of the light illuminating clouds is not direct sunlight but is scattered blue light from the sky. The clouds then appear varying shades of blue-grey, while those that are in complete shadow

SEE ALSO

alpine glow (p. 138)

blue sky (p. 128)

precipitation (p. 146)

purple light (p. 133)

may appear almost black. The color of the base of low clouds is often affected by that of the surface. In arctic regions, snow, ice, and open water may be recognized by subtle differences in the color and brightness of the cloud base.

At sunrise and sunset the colors naturally tend towards yellow, orange and red, and may change dramatically over very short periods, particularly if there are several layers of cloud, which may become prominent in turn as they are successively illuminated by the rising or setting Sun. These conditions often reveal cloud features that otherwise would pass unnoticed. When the purple light is present, some of the clouds may take on an unusual vibrant purple shade.

When there are heavy layers of cloud, the sky beneath them often takes on an orange or reddish tint. This arises for exactly the same reason as sunset colors and the blue sky: the light from the clear sky beyond the clouds has to travel for a considerable distance through the densest layer of the atmosphere, and all the shorter wavelengths are scattered aside, leaving just the long, orange and red wavelengths to reach the observer.

The colors beneath clouds are also affected by precipitation. The color changes are subtle, however, and not always easy to recognize. A yellowish tint is often a warning that rain is imminent, but rain itself produces a reddish shade—again by absorbing the shorter wavelengths. Both hail and snow shafts appear brighter than rain, which normally is light to dark grey in tint. Hail is sometimes reported as producing a greenish shade below the clouds, and hail shafts are (as might be expected) straight, rather than curved like the majority of the trails produced by snow and rain.

This orange, early-morning sky was followed by a day of increasing wind and heavy rain

CREPUSCULAR RAYS

SEE ALSO

blue sky (p. 128)

scattering (p. 128)

sun pillar (p. 122)

Frequently, rays of light or bands of shadow appear to radiate from a single point in the sky. These are crepuscular rays and, despite their name, which associates them with twilight, they may be seen at any time of the day if conditions are appropriate.

There are three slightly different forms of crepuscular rays:

▪ RAYS OF LIGHT THAT PENETRATE THROUGH HOLES IN A LAYER OF CLOUD

▪ RAYS OF LIGHT OR BANDS OF SHADOW THAT RADIATE FROM THE EDGE OF A CLOUD

▪ BANDS OF SHADOW THAT RADIATE FROM BELOW THE HORIZON AT SUNRISE OR SUNSET

All of these types are visible only when the atmosphere contains a plentiful supply of scattering particles, either because of the presence of high humidity or some form of haze. When the atmosphere is perfectly clear, crepuscular rays are invisible.

Broken cloud, such as stratocumulus, is often accompanied by shafts of light, radiating from the position of the Sun, that strike down towards the surface. These are known by a large number of common names, including 'the Sun drawing water', 'Jacob's Ladder', and the nautical 'Apollo's backstays'. They may also accompany shower (cumulonimbus) clouds. In the distance, the darker areas between the rays may appear like precipitation, but this is always unlikely with stratocumulus or other stratiform clouds.

In the second type, pale rays of light or dark bands of shadow fan outwards from the edge of a cloud. It depends on the conditions whether the

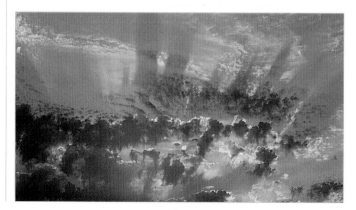

A fine display of crepuscular rays cast by altocumulus castellanus and floccus

rays of light or the shadows are particularly conspicuous. The latter are often prominent with clouds such as altocumulus. At other times, the outline of the cloud may be paralleled by a fringe of shadow. A similar effect sometimes occurs when mountain peaks cast a shadow onto the atmosphere.

Crepuscular rays were originally named from the third type, where long shadows may be cast right across the sky at sunrise or sunset. The bright rays often show a pale reddish tinge and, because of color contrast, the bands of

shadow have a greenish hue. These crepuscular rays are shadows cast by clouds or mountains, far in the distance, sometimes more than 60 mi. away. This type of crepuscular ray generally implies that the sky is clear in the distance and may thus be a useful indicator of forthcoming weather.

Rays that stretch right across the sky and seem to converge at the antisolar point are known as anticrepuscular rays. Sometimes, an isolated cloud may cast a very broad shadow and produce a single anticrepuscular ray, appearing like a V-shaped wedge cut out of a bank of cloud.

Occasionally a gap in the cloud cover allows a shaft of light to illuminate the underside of a layer of clouds. If this appears vertical to the observer, it may be mistaken for a sun pillar. But sun pillars form only when there are ice crystals present, so if the clouds are water-droplet clouds, a sun pillar may be ruled out. Similarly, shadows of distant clouds are sometimes cast upwards onto the base of a cloud layer, producing dramatic bands of shadow.

Crepuscular rays at sunset over Patong Bay, Thailand (TOP)

Crepuscular rays striking through breaks in cloud over Skye (ABOVE)

ALPINE GLOW

A striking sequence of colors is often visible on mountain peaks at sunrise or sunset. At these times also, the solid body of the Earth itself casts a visible shadow onto the atmosphere.

The alpine glow on the Coast Mountains behind Vancouver, British Colombia

As the Sun sinks in the west at sunset, mountain peaks and other objects in the east are bathed in a series of different colors. This is particularly noticeable when the mountain tops are covered in snow. The effect was first generally noticed in the Alps, so it has come to be known as the alpine glow, but is also sometimes called alpenglüh (which is the German form), or alpenglow. The alpine glow is generally strongest when the weather is fine, although it may be quite marked just after rain, when wet rock surfaces may reflect a significant amount of light.

Because the shorter wavelengths of light are scattered by the atmosphere, only the longer wavelengths remain in significant amounts after the sunlight's long path through the atmosphere. The tint of the peaks becomes, successively, yellow, pink, reddish, and finally purple. The last color may seem somewhat surprising, but some scattered blue light does reach the mountain tops at all stages. It is only when the last red rays of light are fading that the blue coloration is sufficiently strong to make a significant contribution, giving rise to the purple shade. When the atmosphere is clear, the colors may be particularly strong and vibrant.

As the Sun sinks still farther, the mountain peaks are overtaken by the shadow of the Earth. Subsequently, if the purple light is present, and when the Sun is about 3–4° below the horizon, the mountains may become tinged with purple once more, where they are reflecting the strong light from the top of the twilight arch. Rather confusingly, this effect is known as the afterglow, which is a term also applied to the last vestiges of light in the twilight arch itself above the Sun.

Although high clouds, particularly dense cumulonimbus clouds, in the east may also display the same sequence of colors, the term alpine glow is usually reserved for the effect on mountain peaks. There is, of course, a similar, reversed sequence of colors at sunrise.

SEE ALSO

blue sky (p. 128)

purple light (p. 133)

twilight arch (p. 132)

SHADOW OF THE EARTH

Like any solid object, the Earth naturally casts a shadow, which falls on the atmosphere, and then stretches out into space. When conditions are favorable, the edge of this shadow may be seen rising in the east as the Sun sinks in the west. It has a distinct color, usually described as steely grey (or blue-grey), and is normally accompanied by a reddish upper border, known as the counterglow. (It is sometimes called the gegenschein, but this term is best reserved for an astronomical phenomenon, a faint glow sometimes visible in the night sky at the antisolar point.) As the shadow rises, it darkens and eventually merges with the dark sky overhead. A similar shadow sinks in the west as the Sun rises in the east at dawn, but this tends to be less distinct, because early in the day the atmosphere is clearer, with less haze particles to render the shadow visible.

The steel-grey Earth shadow, bordered by red, rising in the east as the Sun sank in the west

MOUNTAIN SHADOW

At sunrise or sunset the shadow of a mountain appears as a dark cone stretching away from the observer. If the air is humid or hazy, the shadow may be cast onto the atmosphere itself, like crepuscular rays or the shadow of the Earth, but it then often appears indistinct. It is generally most striking when the atmosphere is clear and the shadow is cast onto the top of a lower layer of clouds.

The shadow always appears as a cone, irrespective of the actual shape of the mountain. In reality, the sides of the shadow are parallel, and their convergence is purely a result of perspective, just like the apparent convergence of railway lines or telephone wires that run into the distance.

When the shadow falls onto cloud, mist or fog, its tip may be surrounded by a glory. The closer the bank of cloud or mist, the more

The shadow of Mt Fuji cast onto a mixed layer of stratus and stratocumulus

Because the Brocken Spectre is an optical illusion, photographs simply show a glory around the shadow of the observer's head.

distinct details of the shadow will become, until eventually it is possible to see the observer's own shadow and that of any companions. Each person will, of course, see his or her own personal glory, not those surrounding the heads of any companions.

The Brocken Spectre

Under certain circumstances, when the cloud is very close or almost enveloping the observer, an optical illusion may come into play. Rather than being cast onto a more-or-less distinct surface, shadows appear at an indefinite distance within the bank of cloud or fog itself, and any variations in the density cause the shadows to appear to approach or recede from the observer. Under these conditions, the brain has difficulty with depth perception. The distance of the shadows is exaggerated, and figures appear to be greatly magnified. This illusion is known as the Brocken Spectre, named after the Brocken, a peak in the Harz Mountains in Germany, from which the effect was first reported.

SEE ALSO

crepuscular rays (p. 136)

glory (p. 106)

heiligenschein (p. 107)

shadow of the Earth (p. 139)

Mist and fog are obscurations of the air caused by suspended water droplets, unlike haze, which is caused by small dry particles. Fog is defined as having a visibility of less than approximately half a mile whereas mist has a visibility greater than that distance. Haze rarely reduces visibility below half a mile.

An early-morning radiation fog

Both mist and fog occur, like cloud, when the air is cooled to the dewpoint. Initially a few, widely spaced, small droplets are formed, producing mist, but if cooling continues these grow in number and size, forming fog. The process is identical to the formation of cloud, and mist and fog may be regarded as cloud at ground-level.

The two most common forms of fog are:

RADIATION FOG: CREATED WHEN THE GROUND RADIATES HEAT AWAY TO SPACE, COOLING THE AIR ABOVE IT

ADVECTION FOG: OCCURS WHEN MOIST AIR IS CARRIED OVER A COOLER SURFACE, REDUCING THE TEMPERATURE TO THE DEWPOINT

RADIATION FOG

There are several conditions that favor the formation of radiation fog (or mist). The sky should be clear, allowing long-wave radiation from the ground to escape into space, and there should be adequate time for temperatures to fall—which favors long winter nights. Obviously the air should be moist during the evening—again this is most likely to occur in autumn and winter. It is also more likely after rain has fallen, in river valleys, or near other bodies of water. Finally, there should be a light wind—no more than about 4 knots (about 4.5 mph)—which ensures that the cooling is spread throughout the lowermost layer of air. This last condition is often found in valleys, and also during anticyclonic conditions.

Fog formation may be extremely rapid when condtions are favorable. Visibility has been known to fall from 2 mi. to less than 650 ft. in just 10 minutes. The depth of this type of fog varies greatly, but is generally between 50 and 300 feet. On rare occasions it may reach as much as 1,000 feet, but this is exceptional. Note, however, that radiation fog does not form over the sea or large bodies of water, because they do not cool rapidly after sunset.

A typical valley fog, as seen from above. The upper surface often shows waves and billows when there is wind shear

Radiation fog disperses when insolation (warming by the Sun) heats the ground after dawn; the ground in turn warming the air above it. Alternatively it may disappear when the wind rises, increasing turbulence, and mixing the fog layer with drier, overlying air. Either process may cause the fog to lift into a layer of low cloud, which may not completely disperse during the day. Frequently, however, when valley fog lifts, it may break up into two lines of fragmentary stratus cloud that hug the valley sides. As heating continues, a valley wind may develop, eventually dispersing the stratus as small clumps of cloud that it carries up the mountain sides.

One particular form of radiation fog is the extremely shallow ground fog (3–6 ft. in depth) that sometimes forms rapidly around sunset, often after late afternoon rain. In this case, it is the lowermost layer of air itself that is cooling by radiation, rather than the underlying ground.

It is possible for water-droplet fog to be supercooled—to exist at temperatures below 32°F. When such a fog drifts across the ground, any of the droplets that come into contact with objects such as vegetation freeze instantly, to give a deposit of rime.

ADVECTION FOG

Unlike radiation fog, advection fog forms over the sea or open ocean. It commonly occurs when warm, moist air flows from a warm surface over cooler water. The temperature of the colder surface must, of course, be below the dewpoint of the air. Another condition for its formation is that there should be a wind, and if the temperature difference is great enough, advection fog may even form under gale-force conditions. Normally, however, strong winds may lead to the formation of low cloud, rather than fog.

The edge of a bank of sea fog being blown onto the coast by the wind

Extensive areas of sea fog may be created in this way, which may invade exposed coasts. Although the fog may burn off inland when the ground warms up though solar heating, along the coast fog may persist throughout the day. After sunset, the fog may spread inland again, assisted by

night-time cooling of the land. Sea fogs that spread inland are sometimes known by the term 'haar', which originated in northeastern Britain.

A similar form of fog may be created when warm, moist air flows over a cold land surface. This often happens when a thaw sets in, because thawing snow causes the ground temperature to hover around 32°F.

OTHER FORMS OF FOG

If cold air flows over warm water, any water vapor arising from the surface condenses immediately into wreaths of steam, giving rise to what is known as a steam fog, often about 20 in. high. This type of fog may be seen frequently above ice-free rivers or lakes on cold winter mornings. Its most dramatic form is, however, seen in polar regions when cold air flows over open water, when it is known as Arctic sea smoke. Although water vapor evaporates into overlying drier air, it may still produce a layer several yards thick.

Smog (smoke fog) forms when the air is full of smoke particles, which act as extremely effective condensation nuclei, and create innumerable tiny condensation droplets. These are smaller than normal fog droplets and form more rapidly with falling temperatures than a pure water fog. The large number of effective nuclei also contribute to smog's persistence. (Note that this form of smog should not be confused with the type that forms in California and elsewhere when ultraviolet radiation converts pollutants from vehicles and other sources into a dense brown haze.)

A form of fog that is rare in western Europe, but quite common in North America, Siberia, and over Antarctica, is ice fog. This arises when temperatures drop so low that fog droplets freeze into tiny ice crystals. These tiny crystals do not reduce visibility, and they glitter in the Sun, giving rise to the popular name of 'diamond dust'. Sunlight refracted through these crystals produces brilliant haloes and other optical phenomena, and some of the rarest types have been observed in this form of fog.

SEE ALSO

anticyclonic conditions (p. 182)

haloes (p. 116)

rime (p. 148)

stratus (p. 26)

supercooling (p. 150)

valley wind (p. 162)

Arctic sea smoke rising from Loch Linnhe in Scotland

Precipitation (which, although related, should not be confused with the supplementary cloud feature praecipitatio) is any form of water or ice that is deposited on the ground or any other surface. It actually includes many different forms, not all of which may be discussed here.

Precipitation may be roughly subdivided into two categories: deposits that form directly on the ground, and particles that fall through the atmosphere.

Precipitation that is found on objects on the ground includes:

- DEW
- GUTTATION DROPS
- HOAR FROST
- RIME
- GLAZE

DEW

Under clear skies, objects rapidly lose heat after sunset through radiation to space. When surfaces exposed to the air, such as blades of grass or leaves, cool below the dewpoint, droplets of water (dew) are deposited on them. Generally, the source of the water vapor is the underlying soil, which, partly insulated by the air lying between the leaves and the ground, remains slightly warmer than higher surfaces.

Heiligenschein around the shadow of the photographer on dew-covered grass

Dewdrops are relatively small, normally less than four-hundreds of an inch in diameter. When deposited on grass they may produce the optical effect known as the heiligenschein, in which a halo of light appears around the shadow of the observer's head. Dewdrops suspended on spiders' webs appear to be particularly effective is producing dewbows.

Not all water droplets seen on plants are dewdrops. All plants transport large quantities of water from their roots to their leaves, where it is evaporated. When the air is very humid, although the water transport continues, the water vapor cannot evaporarte into the air, which is already saturated. The water is exuded as droplets, known as guttation drops, from the very tips of the leaves or blades of grass. These guttation drops are larger than dewdrops, being normally at least eight-hundredths of an inch across. They often display tiny spectra within them, just like that seen in a rainbow.

FROST

Sometimes the temperature will drop below freezing before any dew is deposited. Water vapor then forms ice crystals on exposed surfaces, without going through the liquid phase. The soft, white crystals of ice are known as hoar frost and may sometimes be so thick that the deposit looks like snow.

Hoar frost deposits often form initially on the edges and tips of leaves and other objects, because these have slightly lower temperatures than the rest of the object. Hoar frost should not be mistaken for rime, which forms by a different process.

Occasionally, dew is deposited onto various surfaces and the temperature then drops below freezing. The dewdrops do not necessarily freeze immediately, but exist in a supercooled state below 32°F. This is commonly a stage in the formation of the fern-like patterns of frost seen on windows. Tiny, supercooled droplets occur on the inside of a window pane or on the outside of car windows. Once a few ice crystals have formed, they grow at the expense of the supercooled droplets, which evaporate as the crystals gradually expand in intricate patterns over the whole of the glass. It is often possible to see a narrow gap between the ice crystals and the surrounding supercooled dew droplets.

Supercooled dew will freeze if the temperature drops below approximately 32°F., and this can occur as cooling continues. This frozen dew is sometimes known as white dew or silver frost. Normally the frozen droplets are distinct, but if the dew has been extremely plentiful, it may form a thin layer of ice, which should not be confused with the form of precipitation known as glaze.

Frost patterns on glass formed by supercooled droplets (TOP)

A typical heavy hoar frost deposit created overnight (BOTTOM)

RIME

Rime is a white deposit on exposed objects, which looks superficially like ordinary hoar frost. It forms, however, by a completely different method, being deposited from supercooled fog. The tiny fog droplets remain liquid until they come in contact with any obstacle, when they freeze almost instantaneously.

Long 'feathers' of rough ice build up on the windward surfaces of objects. Although any wind associated with foggy conditions is normally fairly light, if the fog persists large amounts of ice may be deposited. In lowland areas it may amount to about half an inch per day, but in mountainous regions that are enveloped in supercooled cloud for long periods, massive amounts may accumulate, sometimes many yards across. These may break away and fall into valleys below, where they may add significant amounts of ice to valley glaciers. Such large accumulations are also frequently found on tall radio and television masts.

Generally, the 'feathers' point into the wind, but under very still conditions, needle-like crystals may grow around the edges of leaves and similar sharp angles, somewhat resembling frost deposits, but with much longer individual crystals. Large deposits of rime may sometimes cause mechanical damage to trees and shrubs, but this is generally far less severe than that caused by glaze.

A tree with moderate rime accumulation, with altocumulus in the sky above (BELOW)

These rime crystals show that the air drifted upwards from the left, passing up and over the col of the Jungfraujoch in the Alps (BELOW RIGHT)

Glaze is a layer of clear ice that forms when supercooled raindrops fall onto exposed objects whose temperature is below freezing. Although the droplets freeze extremely rapidly, they have sufficient time to spread out into a thin layer before doing so. As a result, the surface becomes coated with a thin layer of ice.

SEE ALSO

condensation
(p. 150)

fog (p. 142)

depression (p. 180)

fronts (p. 179)

dewpoint (p. 187)

heiligenschein
(p. 107)

dewbow (p. 115)

supercooling (p. 150)

The 'black ice' often mentioned in weather forecasts for motorists is a thin, transparent layer of glaze that forms on road surfaces, and which is so dangerous because, visually, the road appears wet, rather than icy.

Glaze frequently forms ahead of an approaching warm front in a depression. When the temperature of the air ahead of the front is below freezing, any rain that falls from the warm, moist air overhead, will be transformed into glaze when it hits the ground. If the front is slow-moving or stationary, thick layers of glaze may accumulate, giving rise to an 'ice storm'. This may be phenomenally destructive, like the ice storm that hit Canada and the northeastern United States in early 1998, breaking down trees, telephone and power lines, and causing electricity pylons to collapse.

A form of glaze may sometimes be created when warm, humid air suddenly brings a thaw to a region that has been undergoing an extremely severe frost. Initially, objects are so cold that water vapor is deposited directly onto surfaces, covering them with a layer of ice. In North America, these conditions occur moderately frequently, giving rise to what is known as a 'silver thaw'.

Utility poles broken by heavy deposits of glaze during the Canadian ice storm of January 1998

CONDENSATION, COLLISION, AND FREEZING

Water vapor cannot condense into droplets without the presence of condensation nuclei. These are, however, so numerous throughout the atmosphere, that when the dewpoint is reached, innumerable tiny cloud droplets are formed. These droplets are very small, ranging from about 1 to 50 μm (i.e., 0.00004–0.002 in.), and growth by condensation is extremely slow, so other processes are required to form raindrops, which are typically 0.02–0.1 in. or more in diameter.

The two dominant processes are growth by freezing, and by collision. Meteorologists sometimes refer to these colloquially as giving rise to 'cold rain' and 'warm rain', respectively. Like condensation, freezing normally requires the presence of suitable nuclei, but these are not always present in large numbers. In the absence of suitable nuclei, cloud droplets may exist at temperatures well below 32°F. Only at about -40°F, will the droplets freeze spontaneously. Supercooling is found in many clouds and also occurs in certain fogs, and the droplets freeze instantly on contact with a suitable nucleus or surface. At high temperatures (about -14°F) relatively few large crystals are formed, whereas at low temperatures (about -22°F) numerous tiny crystals are created. The size of the crystals, which melt when they fall into warmer air, determines whether the resulting raindrops will reach the surface.

In winter, quite shallow cumulus clouds may become glaciated and give rise to a shower, whereas in summer, even towering cumulus congestus may not turn into cumulonimbus, because their tops fail to reach sufficiently low temperatures.

The growth of droplets by collision is actually quite a complicated process in which a number of factors are present. Basically, large drops fall faster than smaller ones, and thus tend to overtake and collide with them, assimilating them and growing as they do so. This process is obviously most likely to occur in very deep clouds, such as the cumulus congestus that are found in summer. Shallow, stratiform clouds are unlikely to produce large raindrops or large amounts of rain, although they do frequently give rise to weak drizzle. It has been established that it may take between 20 and 60 minutes for a droplet to grow to the size of the average raindrop, so clouds with short lifetimes are also unlikely to give appreciable rainfall.

The edge of an approaching heavy rain shower, showing a characteristic reddish tint beneath the cloud (FAR RIGHT, BELOW)

RAIN

Rainfall is conventionally divided into:

- RAIN: DROPS WITH DIAMETERS GENERALLY 0.02–0.1 IN.
- DRIZZLE: DROPS WITH DIAMETERS BELOW 0.02 IN.

The clouds that may produce each of these types of precipitation are:

- RAIN: THICK STRATOCUMULUS (SC STR OP), NIMBOSTRATUS, ALTOSTRATUS, ALTOCUMULUS FLOCCUS, ALTOCUMULUS CASTELLANUS, CUMULUS CONGESTUS, CUMULONIMBUS
- DRIZZLE: THICK STRATOCUMULUS (SC STR OP), STRATUS

> With vigorous cumulonimbus clouds there is a rapid growth of ice crystals, which soon give rise to heavy rainfall

Cumulus congestus and cumulonimbus produce rain in the form of showers, whereas nimbostratus, altostratus, and stratocumulus are the principal rain-bearing clouds in depression systems. Both warm and cold fronts may produce substantial amounts of rain. The quantity of rain that reaches the ground from the altocumulus species is normally very low.

Rain or drizzle may freeze on contact with the cold ground or other surfaces, and is then known as freezing rain (or freezing drizzle). This occurs when the rate of precipitation is low, so the droplets do not have a chance to spread into a layer before freezing, which would otherwise produce a coating of glaze.

There are several different forms of frozen precipitation; the differences in some of these forms lie in the method of their formation and the clouds from which they fall.

- SNOW: LOOSE CLUMPS OF CRYSTALS
- SLEET: MELTING SNOW, OR MIXED RAIN AND SNOW
- SNOW GRAINS: TINY, OPAQUE, WHITE GRAINS OF ICE
- SNOW PELLETS: WHITE, OPAQUE GRAINS OF ICE, GENERALLY SPHERICAL, 0.08–0.2 IN. IN DIAMETER
- ICE PELLETS: TRANSPARENT OR TRANSLUCENT, IRREGULAR OR ROUGHLY SPHERICAL PELLETS OF ICE, LESS THAN 0.2 IN. IN DIAMETER
- SMALL HAIL: TRANSLUCENT, IRREGULAR OR SPHERICAL PELLETS, LESS THAN 0.2 IN. IN DIAMETER
- HAIL: SPHERICAL BALLS OF ICE, 0.2–2 IN. OR MORE IN DIAMETER, SOMETIMES FALLING AS IRREGULAR CLUMPS

SNOW

With rain, the ice crystals that are formed high in the cloud melt on their way down to the surface. The original ice crystals have the intricate, six-sided structures that are conventionally regarded as 'snowflakes'.

Even a modest fall of snow consists of untold millions of individual snowflakes, which themselves consist of clumps of many different ice crystals.

In fact, actual snowflakes consist of innumerable individual crystals that have collided with one another, and become frozen together. This generally occurs in clouds in which the temperature is just below 32°F. At these temperatures, the thin film of water on the surface of the crystals freezes and locks them together. At lower temperatures, the individual crystals remain separate. At any temperature below freezing, either type produces a fall of snow.

The type of snow that reaches the ground is entirely dependent on the temperature. At low temperatures, loose powder snow consisting of small ice crystals results, but close to freezing, large flakes of 'wet snow' will fall. This form of snow often poses most difficulties for transport authorities, because unlike powder snow, which may be blown out of the way, wet snow melts with the slightest pressure, but refreezes as soon as the pressure is removed. It therefore creates hazardous driving conditions and presents serious problems at airports and on railway tracks.

SLEET

In North America, the term sleet is used both for wet snow and also for ice pellets. In Britain, sleet is the accepted term for partially melted snowflakes, or a mixture of rain and snow.

SNOW GRAINS

Snow grains are tiny grains of ice, generally less than 0.04 in. across. This type of precipitation may be regarded as frozen drizzle, and falls from similar stratiform clouds to those that produce this form of precipitation.

SNOW PELLETS

Tiny supercooled droplets freeze together into opaque grains of ice, which appear white because air is trapped within them. They may be regarded as an incipient form of hail, and are sometimes known as soft hail or graupel.

ICE PELLETS

These consist of frozen raindrops or melted and refrozen snowflakes. In both cases the particles either originated in, or passed through a warm layer of air before falling into a colder one in which they froze. They sometimes occur in a conical form.

Fresh snow contains large amounts of air

SMALL HAIL

These translucent pellets consist of a core of snow, surrounded by a layer of clear ice. As for ice pellets, they are normally spherical or irregular, but are sometimes conical in shape. The process by which they form is described next under hail.

HAIL

SEE ALSO

depressions (p. 180)

dewpoint (p. 115)

fog (p. 142)

fronts (p. 179)

showers (p. 165)

supercell storms (p. 169)

Hailstones consist of alternating layers of clear and opaque ice. These layers are laid down in regions at different temperatures: the opaque layers where supercooled droplets freeze on impact, trapping air between them; and the clear layers in warmer regions of the cloud where liquid droplets spread over the surface of the growing hailstones before freezing to give what is, in effect, a layer of glaze. Conditions for this to occur are found in deep cumulonimbus clouds, where the strong updrafts are normally tilted, allowing the growing hailstones to be carried upwards, tossed out of the updraft at a high level to fall back towards the surface, only to encounter the updraft at a lower level and repeat the process. The growing hailstones may circulate several times in this manner, laying down multiple layers of clear and opaque ice. Eventually the hailstones grow too large to be supported by the updraft, and fall out of the cloud.

Although substantial hail may be produced by moderate-sized cumulonimbus clouds, the very largest hailstones are created by supercell storms,

Heavy hail usually precedes any rainfall

Hail falling simultaneously with rain (as shown by the rainbow) over the Solway Firth

Half-inch-sized hailstones, some of which reveal their internal concentric structure

which contain exceptionally strong updrafts. These supercells may produce devastating hail storms, which cause extensive damage to crops, buildings and vehicles, as well as injuring and killing animals and people. They have been known to produce grapefruit-sized hailstones. The largest individual hailstones ever recorded weighed 2.2 lb., but hailstone aggregates, where several have frozen together into a single mass, have reached approximately 9 lb. These records falls came from Bangladesh and China, respectively.

The clouds from which frozen precipitation normally falls are:

- SNOW AND SLEET: NIMBOSTRATUS, ALTOSTRATUS, THICK STRATOCUMULUS (SC STR OP), CUMULONIMBUS
- SNOW GRAINS: THICK STRATOCUMULUS (SC STR OP), STRATUS
- SNOW PELLETS: CUMULONIMBUS (IN WINTER-TIME)
- ICE PELLETS: NIMBOSTRATUS, ALTOSTRATUS, CUMULONIMBUS
- SMALL HAIL: CUMULONIMBUS
- HAIL: CUMULONIMBUS

The type of precipitation known as diamond dust, described under fog, may also be produced under very cold conditions by stratus, nimbostratus, or thick stratocumulus.

It might seem a silly question to ask which way the wind is blowing—after all, anyone can feel the wind on his or her face. But it is not quite as simple as that.

Winds are produced by differences in pressure, themselves caused by differences in temperature. When air is heated, it expands and its pressure becomes less. Conversely, when cooled, it contracts, and its pressure increases. You might think that air would flow directly from high pressure to low pressure, but another force, produced by the Earth's rotation, comes into play, preventing it from doing so. As a result, over most regions of the Earth, air flows round areas of high pressure (anticyclones) and low pressure (cyclones or depressions). This situation corresponds approximately to the flow of air at a moderate height in the atmosphere, away from the turbulence and friction caused by the surface. For practical purposes, this height may be taken as that of low clouds—about 2,000 feet.

The forces act in such a way that, in the northern hemisphere, winds flow clockwise round high-pressure regions, and counterclockwise round low-pressure ones. In the southern hemisphere, directions are reversed. This gives rise to a famous rule of thumb, known as Buys Ballot's law (after the Dutch scientist who described it):

With your back to the wind, low pressure is on your left (in the northern hemisphere).

Because of friction from the surface, however, the balance of forces is altered, and the direction of the surface wind is slightly different from that at low-cloud level. The result is that winds spiral out from high-pressure regions and into low-pressure ones. The change in wind direction depends upon the amount of friction, less over the sea than over the land. Although there is some variation, the change in direction generally amounts to about 10–15° over the sea, and 40–50° over land. So Buys Ballot's law is not completely accurate for the surface wind. In fact, the low-pressure center will not be directly to your left (in the northern hemisphere), but some 10–50° further forward. In the first few hundred yards as one goes higher in the atmosphere, the wind gradually swings round to the freely flowing direction (and also becomes stronger).

This change in wind direction with height often causes confusion. A menacing shower cloud that is directly upwind will actually move to the left (in the northern hemisphere), and may miss the observer completely. One that is (say) 40° to the right of the surface wind may, however, pass directly overhead. It is also likely to arrive slightly sooner than expected, because of the greater wind speed at that level. Shower clouds and larger thunderstorms can cause even more confusion, because they actually create a complicated pattern of winds around them, which we will discuss later.

This image from Meteosat, taken in February, shows prominent depressions in both the northern and southern hemispheres, where air is swirling into low-pressure centers. Although not as readily seen, air is flowing out from the high-pressure regions over the Sahara and South Africa

Two useful terms to describe any change in wind direction are veering and backing. If, when facing the wind, it changes to come from farther to the right it is said to veer. Backing is a change to the left. In the northern hemisphere, a wind that veers moves in the same direction as the Sun. Note that these terms imply the same movement in both hemispheres.

VEER: CHANGE IN A CLOCKWISE DIRECTION, MOVING FARTHER TO THE RIGHT

BACK: CHANGE IN A COUNTERCLOCKWISE DIRECTION, MOVING FARTHER TO THE LEFT

Regardless of whether we are closest to a high-pressure or low-pressure center:

IN THE NORTHERN HEMISPHERE, THE WIND VEERS AND INCREASES WITH HEIGHT

IN THE SOUTHERN HEMISPHERE, THE WIND BACKS AND INCREASES WITH HEIGHT

WIND SPEEDS

For meteorological purposes, the speed of the wind, which is obtained by an instrument called an anemometer is always measured at a standard height of 10 on land. At sea—on ocean buoys, for example—a different height may be used, but the values are adjusted to the standard height. Meteorologists sometimes quote wind-speeds in knots (the standard usage in aviation) or yards per second. In this book, some speeds are given in miles per hour.

A method often used in weather forecasts, especially those intended for sailors, is to quote wind speeds in terms of the Beaufort scale. This was introduced by Rear-Admiral Sir Francis Beaufort in 1806 (when he was still a Commander), and adopted by the British Admiralty in 1838. Although originally described in terms of the sails that could be carried by frigates of the period, the scale was subsequently adapted for use on land, and modified to be of use with any vessel. The scale is defined in knots, so the values in mph given here are approximate equivalents. The table details the Beaufort Scale born for use at sea and on land.

FORCE	DESCRIPTION	SEA STATE	KNOTS	MPH
0		calm like a mirror	below 1	below 1
1	Light air ripples	no foam	1–3	1–3
2	Light breeze	small wavelets with smooth crests	4–6	4–6
3	Gentle breeze	large wavelets; some crests break; a few white horses	7–10	7–11
4	Moderate breeze	small waves; frequent white horses	11–16	12–18
5	Fresh breeze	moderate, fairly long waves; many white horses; some spray	17–21	19–23
6	Strong breeze	some large waves; extensive white foaming crests; some spray	22–27	24–30
7	Near gale	sea heaping up; streaks of foam blowing in the wind	28–33	31–36
8	Gale	fairly long & high waves; crests breaking into spindrift; foam in long prominent streaks	34–40	37–44

FORCE	DESCRIPTION	SEA STATE	KNOTS	MPH
9	Strong gale	high waves; dense foam in wind; wave-crests topple and roll over; spray interferes with visibility	41–47	45–52
10	Storm	very high waves with overhanging crests; dense blowing foam, sea appears white; heavy tumbling sea; poor visibility	48–55	53–61
11	Violent storm	exceptionally high waves may hide small ships; sea covered in long, white patches of foam; waves blown into froth; visibility severely affected	56–63	62–70
12	Hurricane	air filled with foam and spray; extremely bad visibility	≥ 64	≥ 71

FORCE	DESCRIPTION	EVENTS ON LAND	KNOTS	MPH
0	Calm	smoke rises vertically	below 1	below 1
1	Light air	direction of wind shown by smoke, but not by wind-vane	1–3	1–3
2	Light breeze	wind felt on face; leaves rustle; wind-vane turns to wind	4–6	4–6
3	Gentle breeze	leaves and small twigs in motion; wind extends small flags	7–10	7–11
4	Moderate breeze	wind raises dust and loose paper; small branches move	11–16	12–18
5	Fresh breeze	small leafy trees start to sway; wavelets with crests on inland waters	17–21	19–23
6	Strong breeze	large branches in motion; whistling in telephone wires; difficult to use umbrellas	22–27	24–30
7	Near gale	whole trees in motion; difficult to walk against wind	28–33	31–36
8	Gale	twigs break from trees; difficult to walk	34–40	37–44
9	Strong gale	slight structural damage to buildings; chimney pots, tiles, and aerials removed	41–47	45–52
10	Storm	trees uprooted; considerable damage to buildings	48–55	53–61
11	Violent storm	widespread damage to all types of building	56–63	62–70
12	Hurricane	widespread destruction; only specially constructed buildings survive	≥ 64	≥ 71

LOCAL WINDS

There are a number of localized winds that last for part of a day, before dying away or being replaced by a different wind. These winds are created by a particular distribution of land and water, or by differences in elevation. They are:

- SEA BREEZE
- LAND BREEZE
- LAKE BREEZE
- VALLEY WIND
- MOUNTAIN WIND

SEA AND LAND BREEZE

During the daytime, the land warms more rapidly than the sea, heating the air above it. The air rises and expands, reducing pressure of the land, so cooler air moves in from the sea, setting a sea breeze in motion. This usually begins around noon and reaches a maximum strength during the afternoon, but then dies away in the evening. Along the coast, and particularly in spring and early summer, when sea temperatures are low, the onset of a sea breeze may bring a mist in from the sea, so that a bright sunny morning may change into a dull, cool afternoon.

As the afternoon progresses, the cool sea air penetrates farther and farther inland, and may reach several tens of miles from the coast (or even more in certain cases). There is often a marked sea-breeze front, where there is a distinct change in temperature. Air is rising at this front, so this

The sea-breeze front, and its accompanying clouds, moves inland as the day progresses

(LEFT) clouds building up on the sea-breeze front and (BELOW LEFT) a thin sheet of cloud drawn seawards by the returning air about 20 minutes later

is often marked by a line of cumuliform cloud, which may become quite deep if the air is also forced to rise over any hills or mountains that are close to the coast. The arrival of a sea-breeze front may be the trigger for the growth of cumulus congestus or cumulonimbus clouds that give rise to subsequent rainfall.

The direction of a sea breeze tends to be at right-angles to the coast, particularly on otherwise windless days. When the land is in the form of a peninsula, sea breezes often converge from opposite coasts, leading to particularly strong growth of cloud along the center of the peninsula.

The air that rises at the sea-breeze front flows back out to sea at a moderate height in the atmosphere. Sometimes a thin sheet of cloud may be observed spreading back out towards the sea in this higher airflow.

A land breeze is the night-time counterpart of a sea breeze. It develops because the land cools particularly quickly at night (especially in the absence of any cloud cover). The denser, cooler air flows out towards the warmer sea. Such a land breeze generally begins about midnight and continues until dawn. Just as with a sea breeze, a land-breeze front develops along the boundary between the warm and cool air, and this may

SEE ALSO

anticyclonic weather (p. 182)

cumuliform cloud (p. 14)

fronts (p. 179)

mist and fog (p. 142)

stratus (p. 26)

valley fog (p. 142)

create a line of convective cloud that slowly moves out to sea during the night. Such lines of cloud that run parallel to the coastline may frequently be seen in satellite images obtained in the early morning.

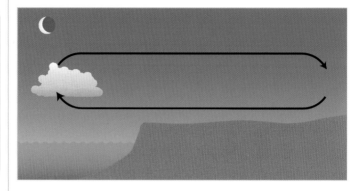

Lake breeze

Large inland bodies of water may produce an effect similar to that of a sea breeze, although here the depth of the water also plays a part. Shallow lakes warm more rapidly than deep ones, reducing the temperature contrast between air over the lake and that over the surrounding land. Large, deep lakes have a much greater effect, and the Great Lakes in North America produce breezes that are similar in strength to sea breezes from the open ocean.

The nature of the surrounding countryside also exerts a strong influence. Shallow lakes set in a flat landscape may produce weak breezes, but if the lake is surrounded by hills or mountains that are strongly heated by the Sun, the lake breeze may be greatly strengthened. Naturally the location of any hills or mountains in relation to the lake also has a considerable influence on the strength of the lake breeze in any particular direction.

VALLEY AND MOUNTAIN WIND

Heating of mountain slopes during the day is also the cause of valley winds. The air rises over the heated slopes and, depending on the orientation of the valley and the degree of heating, creates a wind flowing up the valley, and up the valley sides. If fog has formed in the valley overnight, it often rises and breaks up into patches of stratus cloud that are carried up the mountain sides during the morning.

Valley winds commonly create enough uplift for clouds to form and drift towards the head of the valley or up towards the ridges

A valley wind generally sets in shortly after sunrise, reaches its greatest strength when the slope receives its greatest heating, and dies away just before sunset. Valley winds may reach a speed of 12 mph over sun-warmed slopes, but by contrast over slopes that are not warmed by the Sun, they may be too weak to be readily detected.

The strength of valley winds naturally also depends on the strength and direction of any gradient wind—i.e., that caused by the general pressure distribution. The strongest valley winds develop under the calm, hot conditions found in anticyclonic weather. When there is a brisk gradient wind, the strong turbulence created by rugged country may prevent a true valley wind from becoming established.

A mountain wind is the night-time counterpart of the valley wind. It arises because the high ground radiates heat away rapidly after sunset, cooling the air in contact with it, which, becoming denser, slides down-hill. At the surface, mountain winds tend to be weaker than the corre-

Small cumulus clouds created as air flows up hillsides warmed by the Sun

sponding valley wind, but may reach about 7 mph. The strength may be considerably increased if the wind flows in a confined gorge or canyon.

People sometimes have difficulty in recalling the difference between valley and mountain winds. Remember, however, that all winds are named for their source, so just as easterly winds come from the east, a valley wind rises from the valley, and a mountain wind sinks down from the mountains.

To meteorologists, the word 'shower' has the very specific meaning: a relatively short period of heavy rain from a cumulus congestus or cumulonimbus cloud. This is in contrast to the longer periods of rain, usually of varying intensity, that occur at the fronts in a depression system. Showers may, however, be embedded in frontal cloud, particularly on cold fronts.

The size and duration of the shower, and the intensity and type of precipitation, depend greatly upon the time of year. In the winter, when conditions are generally cold, clouds have a reduced water content, and the freezing level is relatively low. We may expect any precipitation to form by the 'cold' process, i.e., to originate through freezing. Any cumulonimbus clouds are shallow, and the water freezes in the form of innumerable small ice or snow pellets and snowflakes. Depending on the exact temperature profile, these may melt before they reach the surface.

In summer, when temperatures are generally higher, the water content of cumulus congestus and cumulonimbus clouds is much greater and, in addition, the freezing level is at a much higher altitude. Clouds are much deeper than in winter, so during the early part of a cloud's lifetime raindrops are often formed through the 'warm' process, that is, through collision and coalescence. This can occur in a cumulus congestus before it reaches the cumulonimbus stage.

In any shower cloud, the amount of water in the form of raindrops or other precipitation is initially small, and the majority are held sus-

pended by the updraft within the cloud. The largest raindrops tend to break up into smaller droplets, which then grow individually, so the number of droplets increases rapidly. Because the amount of water held in suspension increases, it reduces the strength of the updraft, until eventually the latter collapses, and the rain is released to fall as a heavy shower.

Very deep clouds, such as those found in summer, reach well above the freezing level. Initially, any water droplets are supercooled, and remain liquid. At about 13°F., any freezing nuclei that are present start to act, and some of the droplets freeze into ice crystals. If the cloud reaches still higher levels, the temperature may fall to -40°F., at which supercooled water droplets freeze spontaneously, even in the absence of freezing nuclei. The cloud has become fully glaciated.

Liquid cloud droplets that encounter ice crystals rapidly freeze onto the surface to produce particles of hail. Small hailstones may be swept upwards inside the cloud through regions at different temperatures and gradually become larger and larger. Until eventually they become too heavy to be sustained by the updraft and fall out as hail.

Cumulus congestus and cumulonimbus clouds go through a number of stages in their lifetimes. Generally three stages are recognized:

- EARLY (GROWING) STAGE: CUMULUS CONGESTUS
- MATURE STAGE: THE CLOUD CHANGES TO CUMULONIMBUS CALVUS AND THEN CUMULONIMBUS CAPILLATUS
- LATE (DECAYING) STAGE: GRADUAL DECREASE IN RAINFALL

In general, the first two of these stages each last about 20 minutes. Some large raindrops may fall during the first stage, but most of the precipitation occurs during the mature stage, initially in the form of large raindrops (at the cumulonimbus calvus stage), and then heavy rain that may

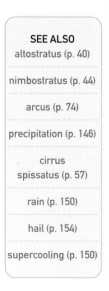

A gust front in the form of a wedge-shaped shelf cloud ahead of an advancing, extremely active cumulonimbus.

SEE ALSO

altostratus (p. 40)

nimbostratus (p. 44)

arcus (p. 74)

precipitation (p. 146)

cirrus
spissatus (p. 57)

rain (p. 150)

hail (p. 154)

supercooling (p. 150)

be accompanied by hail. The last stage may last anything between 30 minutes and about two hours, during which time the rainfall decreases both in intensity and in size of raindrops.

With very light winds, the whole sequence may be observed at a single site, but normally conditions are such that a sequence of showers are experienced at different stages of development. If there is considerable wind shear, with wind speed increasing with height, the top of the cloud may spread out ahead of the main cloud and produce a certain amount of precipitation before the heaviest rain arrives. The overall duration of convective activity and precipitation tends to increase.

If convection is strong enough, a cumulonimbus may reach a stable layer aloft, perhaps even as high as the tropopause. It spreads out beneath the inversion and produces a cloud that could be described as altostratus or nimbostratus, rather than the stratocumulus or altocumulus (Sc cugen or Ac cugen) that arise when ordinary cumulus encounters an inversion. This higher cloud (As cbgen or Ns cbgen) is fibrous in appearance, and is usually accompanied by cirrus spissatus. If there is considerable wind shear, the result is a large overhanging anvil (incus) often with mamma hanging from its underside.

Although each cell has a limited lifetime, after some hours of convective activity, the remnants of several cells may merge into a single, extensive cloud mass. When convection ceases the cumulonimbus starts to decay, leaving behind broken, irregular masses of cloud, at various levels. These are often in the form of cumulus humilis or mediocris, patches of altocumulus and altostratus, and extensive cirrus spissatus. The cirrus may be extremely persistent, and may actually accumulate and last for so long that they inhibit convection the following day.

Any heavy precipitation in the form of rain or hail creates a powerful downdraft of cold air that fans out from the base of the cloud when it hits the ground. The effect of this is most noticeable when the general wind is light. As a shower cloud approaches, the updraft draws air in from around the cloud. As a result an observer will frequently notice a breeze blowing towards the cloud, which will actually be carried along by the wind at low-cloud height. This is the reason for the common misconception that a shower approached 'against the wind'. As the shower comes even closer, the steady inflow is suddenly replaced by a strong, gusty, cold wind blowing out from the cloud.

This cold outflow undercuts the warmer air flowing into the cloud, and the extra lift can initiate the formation of a completely new convection cell, which may itself go on to become a fully-fledged shower. In this way convective activity may gradually become more intense and wide-spread, with many cells at different stages of activity. All will have similar lifetimes, however, although they may become organized into longer-lasting systems, known as multicell and supercell storms, to be described shortly.

When the downdraft is strong, the cold air flowing out at the surface may penetrate many miles ahead of the storm. It may also produce a distinct gust front, with strong unsteady winds, but which is not accompanied by any precipitation. In the larger, more organized systems, the warm air that is undercut by the cold airflow may be lifted above the condensation level, and give rise to a wedge-shaped shelf cloud projecting from the forward edge of the system. This is a specific form of the supplementary cloud feature known as arcus. More rarely, but particularly with the major storm systems, the gust front may be accompanied by a roll cloud, completely detached from the main cloud mass, and which appears to rotate about a horizontal axis.

Decaying cumulonimbus anvils over Lake Baikal

THUNDERSTORMS

Although thunderstorms are (obviously) linked with thunder and lightning, they are also frequently associated with other violent weather phenomena. These include violent squalls, extreme hailstorms, and tornadoes.

The conditions favorable for the formation of thunderstorms are largely the same as those required for the growth of showers. There must be deep convective clouds and a plentiful supply of moisture. In addition, cloud temperatures must be well below -5°F., and both water droplets and ice crystals must be present simultaneously.

Generally, only large cumulonimbus clouds will fulfil these conditions but, occasionally, they may apply in thick, unstable medium-level cloud. Altocumulus floccus and altocumulus castellanus are both indicators of middle-level instability and when deep masses of these clouds are present it is an indication that there may be outbreaks of thundery showers. Such a deep layer of humid air may also trigger explosive growth of cumulus clouds that reach this level, leading to the rapid development of deep cumulonimbus clouds.

Thunderstorms very rarely consist of individual cumulonimbus clouds. In general, they are groups of showers, each of which, as we have seen, has a relatively short lifetime, although the overall group may last for several hours. Sometimes the cells will be relatively disorganized, but more often the inflow into a particularly active cell will lead to the formation of a series of cells, which themselves then grow and continue the storm activity, before decaying in their turn. These are known as multicell storms. At night, it easy to pick out the active cells by the way in which the lightning flashes are grouped in clusters. Although several cells are often active at the same time, it is possible to see how the activity of the oldest cells dies away, and is replaced by lightning from newer cells.

During the daytime, it is often possible to see how the inflow to a

A flanking line lies to the left of the main body of this well-developed cumulonimbus cloud, which later became a full-fledged thunderstorm

major center of activity produces a line of cumulus towers, which generally increase in size towards the main cloud mass. Such a line of cloud is known as a flanking line, and usually lies to the south of the main storm center (in the northern hemisphere).

Occasionally, multicell activity may become organized into a long line extending across country. Humid air being drawn into the storm creates new cells ahead of the line, which grow and become active themselves, causing the line to become self-perpetuating and sweep across the country. Such squall lines may last for many hours.

An even more extreme storm may occur when, instead of being divided into multiple cells, the updrafts and downdrafts, and the consequent activity becomes organized into one giant cell—a supercell. This contains a single, vast rotating updraft (known as a mesocyclone) 2–6 in. across, and a series of strong downdrafts. The circulation within the cloud mass is such that it favors the growth of extremely large hailstones. Tornadoes tend to develop in a region, towards the rear of the storm, known as the rear flank downdraft. Like the multicell storms, supercells are self-perpetuating, and so, unlike cumulus and small cumulonimbus clouds, they persist for many hours or even from one day to the next.

SEE ALSO

altocumulus castellanus (p. 48)

altocumulus floccus (p. 50)

hail (p. 154)

lightning (p. 170)

showers (p. 164)

tornadoes (p. 174)

These cloud-to-ground strokes over East Kilbride, Scotland, were readily visible because cloud base was unusually high for a British storm.

LIGHTNING

Lightning arises when there is an electrical discharge between a cloud and the ground, between clouds, or between different regions of the same cloud. It usually accompanies multicell and supercell storms, but is not, as is commonly believed, always observed before the formation of tornadoes.

Although the exact mechanisms by which thunderstorm clouds accumulate electrical charge remain uncertain, it is believed that the strong updrafts in cumulonimbus clouds serve to sweep particles with positive charge to the top of the cloud, while heavier particles with a negative charge accumulate at the base. This picture is simplistic, however, because pockets of different electrical charge also occur in other regions of the cloud. The negative charge at the base of the cloud induces a positive charge on the ground, which maintains its position beneath the cloud as the latter is carried across country by the wind.

When a sufficiently high charge has accumulated, the air's electrical resistance breaks down, and there is a discharge between the cloud and the ground, usually to the highest point in the immediate vicinity. The discharge actually occurs as a series of strokes, beginning with a stepped leader, which descends from the cloud along a branching path. When this makes contact with a point on the ground the main discharge (or return stroke) flows from the ground, carrying a positive charge up into the cloud. The process may be repeated several times within the few fractions of a second occupied by the overall lightning stroke.

The intense heating caused by the vast electrical current causes the air along the discharge channel to expand and then contract violently, producing the sound of thunder. Timing the interval between seeing the flash and hearing the thunder enables you to make a fairly accurate estimate of the distance to the

A massive cumulonimbus cloud with pileus at sunset, New South Wales. This soon developed into a vigorous, long-lasting thunderstorm.

discharge. Three seconds is approximately equal to half a mile. Thunder is rarely detectable at a distance of more than 18 mi., although under exceptional conditions lightning itself may be seen more than 90 mi. away. Timing the duration of the rumble of thunder will also give an approximation of the length of the discharge channel, which may be many miles long.

Lightning is commonly described as being 'fork' or 'sheet' lightning, but this distinction is rather artificial, because fork lightning is one in which the vertical discharge channel is seen, whereas in sheet lightning it is invisible. The discharge channel is often masked by cloud when flashes occur between different regions of a single cloud, or between clouds. Lightning that is too distant for the thunder to be heard is some-times called 'heat' lightning and thought to occur only when temperatures are high. It is, however, no different from ordinary lightning, and occurs at any time of the year.

There are some unusual (and rare) forms of atmospheric electricity, any observations of which are of great scientific interest:

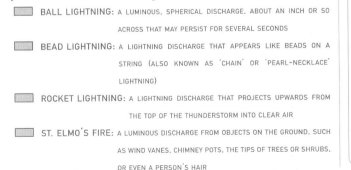

BALL LIGHTNING: A LUMINOUS, SPHERICAL DISCHARGE, ABOUT AN INCH OR SO ACROSS THAT MAY PERSIST FOR SEVERAL SECONDS

BEAD LIGHTNING: A LIGHTNING DISCHARGE THAT APPEARS LIKE BEADS ON A STRING (ALSO KNOWN AS 'CHAIN' OR 'PEARL-NECKLACE' LIGHTNING)

ROCKET LIGHTNING: A LIGHTNING DISCHARGE THAT PROJECTS UPWARDS FROM THE TOP OF THE THUNDERSTORM INTO CLEAR AIR

ST. ELMO'S FIRE: A LUMINOUS DISCHARGE FROM OBJECTS ON THE GROUND, SUCH AS WIND VANES, CHIMNEY POTS, THE TIPS OF TREES OR SHRUBS, OR EVEN A PERSON'S HAIR

SEE ALSO

cumulonimbus (p. 68)

supercell storms (p. 169)

multicell storms (p. 169)

tornadoes (p. 174)

showers (p. 164)

WHIRLS AND DEVILS

There is a whole family of related rotating columns of air, known collectively as whirls or devils. The most familiar is perhaps the dust devil, but there are others, such as water and snow devils.

A well-developed dust devil, photographed in Nigeria

The most common whirls are probably those that are created when funnelling of the wind by surrounding obstacles sets a column of air into rotation. This is exactly the same effect as that seen in cities with high buildings, where the wind howls between the buildings and whips a column of leaves and paper up into the sky. In the countryside, wind blowing through a gap between hills can create a similar column of snow, dust, or (if it crosses a lake) water. Such whirls often remain more-or-less stationary, but are sometimes seen to travel short distances. They may also weaken and reform in roughly the same position all the time that the wind remains in the same direction and at the same strength.

True dust devils occur under hot conditions, and arise when there is intense heating of the surface, producing a rising column of air, whose initial rotation may be induced by irregularities in the surface. Unlike waterspouts, landspouts and tornadoes they grow upwards from the surface, rather than down from the clouds. These devils generally travel across the ground and it is not uncommon to see several at the same time. Apart from dust, they may carry other loose material, such as hay, up from the surface. On rare occasions, a small cumulus cloud may be seen at the top of the rising column of warm air. Only rarely are these whirls strong enough to cause significant damage.

WATERSPOUTS AND LANDSPOUTS

Stronger and longer-lasting than the small whirls and devils are waterspouts and landspouts. The latter term has begun to be employed only recently, but serves to distinguish them from the much more violent tornadoes.

Waterspouts and landspouts originate when strong convection occurs in cumuliform clouds. A vigorous downdraft creates a column of air which grows downwards from the cloud. Because pressure is reduced in the center of the rotating column, condensation is usually present, so the column appears opaque, ranging from white to dark grey, depending on the lighting. The central downdraft is surrounded by a largely invisible updraft.

Frequently these funnel clouds (tuba) merely hang from the base of the clouds and fail to reach the ground. They are commonly seen below cumulus congestus or cumulonimbus clouds. They have been recorded beneath stratocumulus, but probably originated in deeper convective clouds that were hidden by the stratocumulus layer. Their lifetimes are relatively short, rarely more than a few minutes.

If a funnel reaches the sea, it becomes a true waterspout. There is a flattened center (the 'dark spot'), where the central downdraft hits the water. This may be surrounded by a curtain of spray, called the 'bush'. Multiple waterspouts are not uncommon, but all are relatively weak features and the majority dissipate when they cross onto land.

Landspouts have a similar origin and structure, and raise a curtain of fairly light debris, corresponding to the bush in a waterspout. The lifetime of both types is normally about 15 minutes, during which time they may move across the surface at speeds of 9–15 mph. Longer lifetimes and higher speeds (up to 30 mph) have been recorded.

A waterspout beneath a cumulus congestus or small cumulonimbus cloud embedded in stratocumulus, observed from Portland, Dorset

TORNADOES

Tornadoes are by far the most extreme of the family of whirls, and are often highly destructive. Although hurricanes (tropical cyclones) affect far greater areas, the damage caused by tornadoes is highly concentrated, and generally more severe.

Many of the tornadoes and 'twisters' mentioned in the media are actually the less extreme, but still potentially damaging, landspouts (or waterspouts that have come ashore), or else the similar vortices that are often generated by the gust fronts present in powerful thunderstorm systems or on active cold fronts. (Both of these types are sometimes called 'gust-nadoes'.) Hurricanes and other tropical cyclones also generate many tornadoes of this general sort.

True tornadoes are created by massive supercell storms. In these systems the circulation of air becomes organized in such a way that it favors the development of tornadoes. The factors governing their formation are complex, but include the presence of wind shear at middle levels and also the development of a 'mesocyclone': a large-scale system of rotation of the air within the cloud mass, with resulting violent updrafts and downdrafts.

Tornadoes themselves are often preceded by a 'lowering', a relatively restricted area that grows down from the base of the supercell storm. This lowering is generally rain-free, and may develop into a cylinder of cloud (a 'wall cloud'), which exhibits obvious rotation and vertical motion (ascent) of the air within it. These are indications that a tornado is imminent. The main funnel cloud may descend either from a rain-free base or wall cloud. Only when the vortex reaches the ground and raises a debris cloud of material from the surface, is it formally defined as being a tornado—even if the vortex itself is invisible.

A broad tornado funnel in the American Midwest. Little debris is visible, but the tornado was at its most destructive phase

Tornadoes do not normally persist for very long: 15 minutes is a typical lifetime. Some long-duration events are thought to have been caused by a succession of individual tornadoes that formed and decayed, rather than by a single, persistent vortex. The diameter at the base varies considerably, ranging from about 300 to 6,600 feet, with the larger tornadoes generally being the most destructive. These large tornadoes sometimes develop a series of smaller whirls that circulate around the circumference of the main funnel. Path lengths are typically 6–60 mi., and wind speeds within the vortex itself 90–180 mph. (The highest reliably known wind speed is 307 mph, recorded in the Oklahoma City tornado of 1999.)

As with waterspouts and landspouts, the pressure is reduced in the center of the vortex—sometimes by as much as 250 mPa—which causes condensation to occur, thus giving rise to the visible funnel. Initially, the funnel may be relatively short and straight, but although the wind speeds may both strengthen and weaken over the tornado's lifetime, a stage is reached when the funnel begins to elongate and narrow. Once this 'rope' stage occurs, the tornado generally weakens and dissipates. (A similar process is observed in the weaker waterspouts and landspouts.)

This waterspout (which developed from the tuba on p. 79) was extremely weak when compared with the destructive tornado shown opposite

As with other violent phenomena, such as flooding, severe thunderstorms and hurricanes, meteorological services alert the general public in two stages:

WATCH: CONDITIONS ARE SUITABLE FOR THE FORMATION OF TORNADOES; TAKE EARLY
PRECAUTIONS AND ENSURE THAT YOU ARE ABLE TO RECEIVE FURTHER NOTICES

WARNING: A TORNADO IN YOUR AREA IS EXTREMELY LIKELY OR HAS ACTUALLY BEEN
SIGHTED; TAKE COVER IMMEDIATELY IN A PROPER TORNADO SHELTER

SEE ALSO
cumulonimbus
(p. 68)

supercell storm
(p. 169)

thunderstorm
(p. 168)

Some knowledge of the overall circulation in the atmosphere is useful in understanding the weather, and in particular how it changes with time. We need to describe only the broad details here, however.

The general circulation is driven by the imbalance between heating in the equatorial region and cooling at the poles. Warm air rising in the tropics descends in regions that are known as the subtropical anti-cyclones (high-pressure areas), which are centered very approximately at 30° N and S. From here, air at the surface flows both back towards the tropics—creating the trade wind belts—and also farther towards the poles. On the poleward side of the subtropical anticyclones lie belts of mainly westerly winds. These lie approximately between latitudes 40 and 70° N and S, although they vary considerably in strength and position (particularly in the northern hemisphere). These westerlies are extremely important for the weather in temperate regions, such as Europe and North America, because the dominant winds in these zones are from the west or southwest, and carry weather systems eastwards around the globe. Although the winds around individual weather systems (depressions and anticyclones) may be in almost any direction, the systems themselves (and particularly depressions) move from west to east, following the general flow.

Cold, dense air flows out from the poles toward the equator, creating generally easterly winds at the surface. There is an extremely important atmospheric boundary where this cold air meets the warmer, westerly winds that encircle the globe at middle latitudes. There is a sharp temperature contrast across this boundary, which is known as the Polar Front. The two polar fronts (one in each hemisphere) do not run round the Earth at fixed latitudes. Instead, each polar front exhibits a series of irregular and unstable waves, or lobes, around the globe (usually four or five in number), where warm moist air extends towards the poles, and colder air pushes down towards the equator. It is at this boundary that vast swirling low-pressure areas (depressions) are created, where warm and cold air mix, and most of the warmth from the tropics is transferred towards the poles.

The hot, moist air rising over the tropics produces deep cumulonimbus clouds and plentiful rain, particularly in the afternoons, when severe thunderstorms are a regular daily occurrence. Satellite images generally show a line of such convective clouds where the northeast and

(TOP) A typical August Meteosat image, with a well-defined Intertropical Convergence Zone, convective activity over equatorial Africa, and generally clear skies over the subtropical high pressure areas

(BOTTOM) A northern winter Meteosat image, with the polar front dipping well south over North America and the central Atlantic, shown by the long, trailing cold fronts. There is major thunderstorm activity over South America

southeast trade winds converge at what is known as the Intertropical Convergence Zone (ITCZ).

When the air descends in the subtropical anticyclones, the air is compressed and warmed, leading to clear skies and intense solar radiation. This is particularly noticeable in the northern hemisphere, where the Sahara and Arabian deserts, and also the southwest of the United States are often completely cloud-free.

Farther towards the poles, the cloud cover is far more variable, because it is here that successions of depressions march around the globe, interrupted occasionally by extensions of the subtropical anticyclones or other areas of high pressure with relatively clear skies.

The depth of the troposphere, in which most clouds and weather systems occur, varies from equator to poles. The tropopause, the upper boundary of the troposphere, is highest over the tropics and lowest over the poles. There are generally distinct breaks in its level above the subtropical anticyclones and also near the polar fronts. It is at these breaks, where there are great temperature contrasts that the subtropical and polar-front jet streams are located. These (particularly the latter) have a great effect on the growth, movement and decay of weather systems in the layer below them.

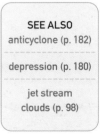

SEE ALSO
anticyclone (p. 182)

depression (p. 180)

jet stream
clouds (p. 98)

AIR MASSES AND FRONTS

When air remains over a particular region for a long period of time, it takes on specific characteristics governed by the nature of the region. It initially retains these characteristics as it moves out from the source region, but becomes increasingly modified with time and distance, depending on the surface over which it is moving. Initially, its temperature and humidity become essentially constant over an areas of thousands of square miles. Temperature is governed by the location of the region, and humidity by whether the air has stagnated over, or passed across, the sea or land. The main sources are the polar regions, the subtropical anticyclones, and the equatorial zone. These are recognized in the classification:

- ARCTIC AND ANTARCTIC (A)
- POLAR (P)
- TROPICAL (T)
- AND EQUATORIAL (E)

If an air mass originates or follows a track over an ocean it becomes humid and is designated 'maritime' (m). If it originates over a large land area, such as the interior of a continent, it tends to remain dry and is called 'continental' (c). The commonest forms of air mass are:

- MARITIME ARCTIC AIR — (MA), VERY COLD AND HUMID
- MARITIME POLAR AIR — (MP), COLD AND HUMID
- CONTINENTAL POLAR AIR — (CP), COLD AND DRY
- MARITIME TROPICAL AIR — (MT), WARM AND HUMID
- CONTINENTAL TROPICAL AIR — (CT), HOT AND DRY

Equatorial air is always maritime in nature (mE), while Arctic or Antarctic air is normally exceptionally cold. The type of clouds that are formed often depends largely on the relative temperatures of the air, as it moves out from its source region, and the underlying surface. Maritime tropical air, for example, that moves over a progressively cooler sea is cooled, and produces sea fogs, and low stratiform cloud, such as stratus and stratocumulus. Such conditions often occur in the warm sector of depressions. Cold polar air or frigid arctic air that moves over a warmer sea becomes highly unstable, producing a host of cumulonimbus clouds and the resulting showers. This type of weather is often found behind the cold front of a depression.

SEE ALSO

anticyclone (p. 180)

depression (p. 180)

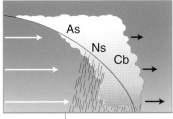

An idealized warm front (ABOVE), showing the succession of cloud types; an idealized cold front (ABOVE RIGHT) as found in many depressions

Fronts

Where two air masses of different types come into contact, the boundary between them is known as a front. Because the two air masses have different temperatures and humidities (and hence densities), there is always a tendency for one (the cooler and denser) to undercut the other (the warmer and lighter). Although fronts may sometimes remain essentially stationary for a period, any small perturbation may destabilize the situation and set both air masses into motion.

At the Polar Front, it is this type of disturbance that initiates the formation of a depression. From a quasi-stationary state, with cold polar air on one side of the front, and warmer tropical air on the other, a wave begins to grow. On the forward (eastern) side of the wave, the cooler air starts to retreat towards the pole, while the warm air advances, and simultaneously begins to slide up the inclined frontal surface. A warm front has come into being.

To the rear of the wave (on the west), the cold air starts to undercut the warm air, lifting it away from the surface. Here, the frontal zone also begins to advance towards the east. A cold front has been created. The formation of a warm front and a cold front is the first stage in the development and evolution of a depression.

Cold fronts frequently occur without any accompanying warm front when cold polar or arctic air sweeps down towards the equator, encountering warmer tropical air as it does so. When there is a large temperature contrast, such cold fronts may develop into vigorous squall lines, with strong convective activity in the form of cumulonimbus clouds, and produce extremely violent weather.

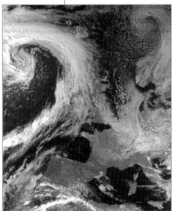

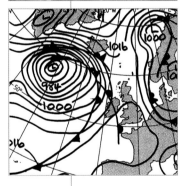

The photograph, taken later than the chart, shows a weak warm front, stretching from Scotland over the North Sea, and a long, trailing cold front

DEPRESSIONS AND ANTICYCLONES

The atmospheric systems that have most influence on the weather are low- and high-pressure regions (depressions and anticyclones). These are most readily seen on the pressure charts—known technically as isobaric charts—that are frequently part of television and newspaper weather forecasts, and which are nowadays also available by fax and over the Internet.

DEPRESSIONS

A depression arises when a closed circulation forms around a low-pressure center (L), with relatively higher pressure (H) around the low. The shaded area indicates the region of general rainfall

On a weather chart, a depression appears as a set of closed isobars that encircle a low-pressure center. Lows extend their influence in the form of troughs, where the isobars form a 'V'-shape pointing away from the center. When it first forms, a depression normally exhibits two troughs, and these lie at the location of the warm front and the cold front. Here the isobars show a distinct change in direction.

The isobars give a fairly accurate representation of the wind direction, at low-cloud level (about 2,000 ft.), which is counterclockwise around the depression center in the northern hemisphere, and clockwise in the southern. As we have discussed earlier, the surface winds spiral into the center, at an angle of 10–50° to the isobars. Because, at the surface, air is converging in the center of the low from various directions, it escapes by ascending, and then flowing outward at a higher level.

All depressions differ to some extent, but they also have certain features in common. The warm and cold fronts, for example, represent the positions at which the warm and cold air, respectively, are advancing. (If either front were stationary, it would be marked as

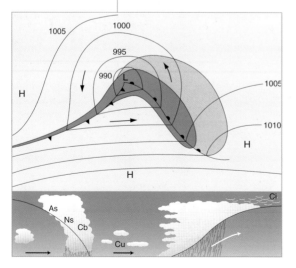

such on the charts.) Between the two fronts is an area of warm air, known as the warm sector.

In the 'classic' model of a depression, the warm front indicates where the warm air is sliding up an inclined surface above the cooler surface air ahead of the depression. As the air rises, it reaches its condensation point and clouds are formed. A similar situation applies at the cold front, where the cold air is undercutting the warmer air. This leads to a particular sequence of clouds as the fronts and warm sector pass over the observer. Ahead of a warm front:

- HIGH CIRRUS THICKENS TO CIRROSTRATUS
- CIRROSTRATUS THICKENS AND LOWERS TO ALTOSTRATUS
- ALTOSTRATUS THICKENS AND LOWERS TO NIMBOSTRATUS, BRINGING MORE-OR-LESS CONTINUOUS RAIN

SEE ALSO

radiation fog (p. 142)

winds (p. 157)

As the warm front passes, the rain slackens, the wind veers, and the cloud tends to become stratocumulus and stratus. If the observer is well away from the center of the low, the cloud cover may break to reveal remnants of the frontal cloud at various levels, such as patches of altostratus and cirrus, as well as some lower cumulus. With broken low cloud, it may be possible to see the succession as the cold front approaches, but this is normally masked by the low stratus or stratocumulus. At a cold front, the clouds are similar to those at a warm front, but roughly in reverse order. The front is also steeper, so they succeed one another more rapidly:

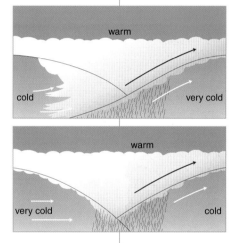

- STRATUS AND STRATOCUMULUS MERGE INTO NIMBOSTRATUS
- ALTOCUMULUS AND ALTOSTRATUS ABOVE THE NIMBOSTRATUS
- CIRRUS AND CIRROSTRATUS TOWARDS THE REAR OF THE FRONT
- CUMULIFORM CLOUDS AND SHOWERS BEHIND THE FRONT

Occluded fronts may be of the warm (TOP) or cold (BOTTOM) type

The high cirrus above the frontal zone is often readily visible once the cold front has passed, and the clearer, cold air has arrived. Quite frequently, there is considerable instability at the cold front, which may then have large cumulonimbus embedded within it. Again, their anvils may be readily visible from the rear of the front. As at the warm front, the wind veers as the front passes.

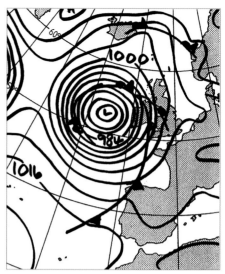

In maritime areas, such as western Europe and western North America, where the air in the warm sector tends to be stable, air is often descending at middle levels, and the fronts are very subdued and ill-defined. Precipitation is light, and the succession of clouds tends to be:

▨ CUMULUS BECOMING STRATOCUMULUS AHEAD OF THE WARM FRONT

▨ THICK STRATOCUMULUS (ST OP) AND LIGHT RAIN AT THE FRONT

▨ STRATOCUMULUS AND STRATUS BEHIND THE WARM FRONT

▨ THICK STRATOCUMULUS AGAIN AT THE COLD FRONT

▨ CUMULUS AND CUMULONIMBUS BEHIND THE COLD FRONT

As a depression evolves, the cold front catches up with the warm one and starts to lift the tip of the warm sector away from the surface. It is at this stage that the winds tend to be strongest. The combined fronts are known as an occluded front, and have a single, merged cloud mass, which often produces heavy rain. The occluded front grows as the depression ages and usually starts to trail behind the low center. Eventually the system and the clouds associated with it dissipate and fade away.

ANTICYCLONES

On a weather chart anticyclones appear as closed isobars around a high-pressure center. In this case, they extend their influence in the form of ridges, which are areas where the isobars are 'V'-shaped, and pointing away from the high-pressure center. This time, however, the surface winds spiral outwards from the center, clockwise in the northern hemisphere, and counterclockwise in the southern.

The weather that accompanies anticyclones tends to be quieter and subject to less dramatic changes than that found with depressions. Because air pressure is high, caused by descending air, cloud growth is restricted or may be entirely suppressed. Skies may be completely clear, but in summer, despite the strong heating of the ground, convection is restricted and any cumuliform cloud may spread out at an inversion created by the subsiding air. Some thin higher cloud, such as disorganized cirrus may also be present. In winter, when conditions are generally colder and the air tends to be more humid, extensive stratus and stratocumulus may occur and persist, giving rise to what is known as 'anticyclonic gloom'. The generally light or non-existent cloud cover leads to low temperatures at night, so anticyclonic conditions favor the formation of radiation fogs, particularly in autumn and early winter.

An anticyclone, centered over the English Channel is giving clear skies over most of Iberia and France, and is blocking the eastward motion of a complex low-pressure area west of Ireland

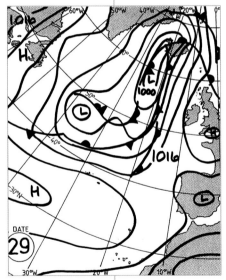

Anticyclones are generally slow-moving and may remain stationary for long periods. This may lead to a 'blocking' situation, where the high pressure forces the depression to follow tracks to north or south of the anticyclone. Consequently, the weather may remain the same for weeks on end. In winter, a blocking anticyclone centered over Scandinavia, for example, may bring icy winds from Russian and W Siberia to W Europe.

NOTE: ALL CHARTS ILLUSTRATE POSITION OF PARTICULAR FRONTS AT 12 NOON

CLOUDS AND WEATHER

The weather is so complex that compiling a detailed forecast requires information about conditions at various levels of the atmosphere and over a large area of the globe. Nevertheless, some useful indications of possible developments may be gained from observing the clouds and winds. If possible, look at the sky several times a day, because this will help you to judge how the situation is developing, and you can modify your tentative forecast accordingly.

Because the actual weather at any particular site is strongly influenced by local conditions—whether to windward or leeward of hills, for example, or proximity to the sea—the information in the following table can serve only as a rough guide to what may be expected.

CLOUDS	PAGE	DESCRIPTION	FORTHCOMING WEATHER
cumulus humilis	23	scattered and moderate in size	fair weather, unless deeper cumulus congestus or high cirrostratus appear
cumulus humilis and cirrus	23, 54	cumulus fail to grow, cirrus increases	warm front approaching, especially if cirrus turns into cirrostratus; rain probably within 12 hours
cumulus congestus	24	towering cumulus	showers possible within the next few hours
cumulus spreading into stratocumulus or altocumulus	87	decreasing clear sky	precipitation unlikely, unless convection strong enough to break through layer; overcast likely to be slow to clear
cirrus	54	high thin clouds, few in number	generally fair, but weather may deteriorate if clouds increase and become organized
cirrus thickening to cirrostratus	61	halo may appear around Sun or Moon	frontal system is approaching with deteriorating weather; rain likely within 6–12 hours
cirrus or cirrostratus and altostratus	101	cloud obviously increasing; strong upper wind spreads any contrails	warm front approaching, wind likely to increase, rain within 6–12 hours
jet-stream cirrus	98	high, fast-moving bands of cirrus often with wave structure	a vigorous depression exists upwind and is approaching; cloud cover will increase; and strong winds are likely in 10–15 hours

CLOUDS	PAGE	DESCRIPTION	FORTHCOMING WEATHER
contrails	100	disappear rapidly	fine, major change unlikely within 24 hours, unless conditions conducive to major convective activity
contrails spreading	100	trails spread, become glaciated and persistent	increasing humidity at height; possible indication of approaching warm front with deteriorating weather
altocumulus floccus or castellanus	49, 50	clumps of altocumulus (often with virga) or lines of towers	thundery showers, possibly severe, likely within 24 hours
altocumulus and cirrocumulus	64	increasing in extent, becoming altostratus and cirrostratus	rapidly deteriorating weather rain within 6–12 hours
altostratus and pannus	40	sky covered with altostratus, ragged pannus increasing	warm front nearby; rain imminent, starting intermittently, but becoming essentially continuous; wind likely to increase
nimbostratus	44	sky completely overcast	more-or-less continuous rain lasting for several hours
cumulonimbus	165	individual clouds of limited horizontal extent	light to moderate showers, gusts; active lifetime limited to 20–30 minutes
cumulonimbus	69	large clouds with several cells and/or anvils	heavy showers with possibility of hail and lightning; strong gusts; lifetime perhaps as long as a few hours
cumulonimbus	165	organized lines or clusters of massive clouds	extremely heavy showers, beginning with severe hail; if part of a cold front will be followed by scattered (and possibly) heavy showers
stratocumulus	32	low cloud with occasional breaks	no significant precipitation; likely to be slow to clear
mixed, stable clouds	38	stratocumulus, altostratus and cirrostratus patches warm sector cloud	no significant change imminent and may persist for days
stratus	26	low cloud	no significant precipitation; if originated as fog, will probably clear later in day, otherwise persistent

FURTHER INFORMATION

Books

Brettle, M. & Smith, B. (1999), *Weather to Sail*, Crowood Press

Chaboud, R. (1996), *How Weather Works*, Thames & Hudson

Dunlop, S. (1999), *Collins Gem Weather*, HarperCollins

Dunlop, S. (2001), *Dictionary of Weather*, Oxford University Press

Eden, P. (1995), *Weatherwise*, Macmillan

File, D. (1996), *Weather Facts*, Oxford University Press

Harding, M. (1998), *Weather to Travel*, Tomorrow's Guides

Ludlum, D.M. (2001), *Collins Wildlife Trust Guide Weather*, HarperCollins

Pedgley, D. (1980), *Mountain Weather*, Cicerone Press

Meteorological Office (1982), *Cloud Types for Observers*, Stationery Office

Watts, A. (2000), *Instant Weather Forecasting*, Adlard Coles Nautical

Watts, A. (2001), *Instant Wind Forecasting*, Adlard Coles Nautical

Whitaker, R., ed. (1996), *Weather: The Ultimate Guide to the Elements*, HarperCollins

Journals

Weather, Royal Meteorological Society, 104 Oxford Road, Reading, Berks. RG1 7LJ (http://www.royal-met-soc.org.uk/weather.html). Monthly magazine

Weatherwise, Heldref Publications, 1319 18th Street NW / Washington, D.C. 20036-1802 (http://www.weatherwise.org/). Bi-monthly magazine

Internet sources

Please note that the URLs may change. If so, use a good search site, such as Google, to locate the information source.

Current weather

BBC Weather (http://www.bbc.co.uk/weather)

CNN Weather (http://www.cnn.com/WEATHER/index.html)

ITV Weather (http://www.itv-weather.co.uk/)

The Weather Channel (http://www.weather.com/twc/homepage.twc)

Wetterzentrale (http://www.wetterzentrale.de/pics/Rgbsyn.gif). Current British station plots; the site itself is in German

General information

NOAA National Climate Data Center (http://www.ncdc.noaa.gov/) National Oceanic and Atmospheric Administration's data site

Reading University (http://www.met.reading.ac.uk/~brugge/index.html). Vast source of data and links compiled by Roger Brugge

UK Weather Information Site (http://www.weather.org.uk/)

WorldClimate (http://www.worldclimate.com/worldclimate/index.htm) Historical data for weather stations around the world

Meteorological offices, agencies and organizations

Environment Canada (http://www.msc-smc.ec.gc.ca/)

ECMWF, European Center for Medium-Range Weather Forcasting (http://www.ecmwf.int/)

Meteorological Office, UK
(http://www.metoffice.com/) Large amount of
information, including useful links
NOAA, National Oceanic and Atmospheric
Administration (http://www.noaa.gov/)
NWS, National Weather Service
(http://www.nws.noaa.gov/)
World Meteorological Organization
(http://www.wmo.ch/)

Societies

American Meteorological Society
(http://www.ametsoc.org/AMS)
Canadian Meteorological and Oceanographic
Society (http://www.meds.dfo.ca/cmos/)
European Meteorological Society
(http://www.emetsoc.org/)
Irish Meteorological Society
(http://homepage.eircom.net/~kcommins/Met
Soc/)
National Weather Association, USA
(http://www.nwas.org/)

Royal Meteorological Society
(http://www.royal-met-soc.org.uk/)
TORRO: Tornado and Storm Research
Organization (http://www.torro.org.uk/)
British amateur-based research organization

Satellite images

Eumetsat (http://www.eumetsat.de/)
Meteosat geostationary satellite images
Remote Imaging Group
(http://www.rig.org.uk/)
Amateur group devoted to reception of satellite
images, at all levels of experience, and with both
inexpensive and more sophisticated equipment
University of Dundee
(http://www.sat.dundee.ac.uk/)
Current and archive of polar-orbiting satellite
images of UK and Europe. Registration (free)
required
University of Strasbourg (http://www-grtr.
u-strasbg.fr/) High-resolution polar-orbiter
images and other material

GLOSSARY

anticyclone A high-pressure region which is a source of air that has subsided from higher altitudes, and from which air flows out over the surrounding area. The circulation around an anticyclone is clockwise in the northern hemisphere.

antisolar point The point on the sky directly opposite the location of the Sun.

backing A counterclockwise change in the wind direction, i.e., from West, through South, to East.

col An area of slack atmospheric pressure, located between a pair of low-pressure centers and a pair of high-pressure ones. Slight changes in pressure may cause a col to move rapidly or disappear.

continental climate A climate that is typical of continental interiors, and characterized by extremely cold winters and hot summers. There is also a tendency towards low overall precipitation totals.

convection Transfer of heat by the motion of parcels of a fluid such as air or water. In the atmosphere this motion is predominantly vertical. There are two forms of convection: 'natural convection' in which parcels or 'bubbles' of air are free to move vertically driven by buoyancy effects; and 'forced convection' in which air is mixed mechanically by eddies.

Coriolis force The apparent force that deflects any moving object (such as a parcel of air) away from a straight-line path. In the northern hemisphere it acts towards the right, and in the southern, to the left. It increases in proportion to the velocity of the moving object.

cyclone A system in which air circulates around a low-pressure core, with two distinct meanings: 1) a 'tropical cyclone', a self-sustaining tropical storm, also known as a hurricane or typhoon; 2) an 'extratropical cyclone' or depression, a low-pressure area, which is one of the principal weather systems in temperate regions.

cyclonic Moving or curving in the same direction as air that flows around a cyclone, i.e., counterclockwise in the northern hemisphere, clockwise in the southern.

depression The most frequently used term for a low-pressure area. Air flows into a depresssion and rises in its center. Known technically as an 'extratropical cyclone'. The wind circulation around a depression is cyclonic (counterclockwise in the northern hemisphere).

dewpoint The temperature at which a particular parcel of air, with a specific humidity, will reach saturation. At the dewpoint, water vapor will begin to condense into droplets, giving rise to a cloud, mist or fog, or depositing dew on the ground.

hurricane One of several names for a potentially destructive tropical cyclone, used in the North Atlantic and eastern Pacific.

instability The condition under which a parcel of air, if displaced upwards or downwards, tends to continue (or even accelerate) its motion. The opposite is stability.

inversion An atmospheric layer in which temperature increases with height.

isobar A line that joins points on a weather chart that have the same barometric pressure.

jet stream A narrow ribbon of high-speed winds that lies close to a break in the level of the tropopause, with two main jet streams (the polar-front and sub-tropical jet streams) in each hemisphere. Other jet streams exist in the tropics and at higher altitudes.

lapse rate The rate at which temperature changes with increasing height. By convention, the lapse rate is positive when the temperature decreases, and negative when it increases with height.

latent heat The heat that is released when water vapor condenses or freezes into ice crystals. It is the heat that was originally required for the process of evaporation or melting.

maritime climate A climate that is strongly influenced by the region's proximity to the ocean. Generally characterized by significant amounts of precipitation throughout the year, but with generally mild winters and summers that rarely reach extremely high temperatures.

mesosphere The atmospheric layer above the stratosphere, in which temperature decreases with height, reaching the atmospheric minimum at the mesopause, at an altitude of either 52 or 60 mi. (depending on season and latitude).

parhelion The technical term for a mock sun.

precipitation The technical term for water in any liquid or solid form that is deposited from the atmosphere, and which falls to the ground. It excludes cloud droplets, mist, fog, dew, frost and rime, as well as virga.

pressure tendency The change in atmospheric pressure during the previous three hours.

ridge The extension of an area of high pressure, resulting in approximately 'V'-shaped isobars pointing away from the pressure center.

stratosphere The second major atmospheric layer from the ground, in which temperature initially remains constant, but then increases with height. It lies between the troposphere and the mesosphere, with lower and upper boundaries of approximately 5–12 mi. (depending on latitude) and 30 mi., respectively.

stability The condition under which a parcel of air, if displaced upwards or downwards, seeks to return to its original position rather than continuing its motion.

supercooling The conditions under which water may exist in a liquid state, despite being at a temperature below 32°F. This occurs frequently in the atmosphere, often in the absence of suitable freezing nuclei.

synoptic chart A chart showing the values of a given property (such as temperature, pressure, humidity, etc.) prevailing at different observing sites at a specific time.

thermal A rising bubble of air, which has broken away from the heated surface of the ground. Depending on circumstances, a thermal may rise until it reaches the condensation level, at which its water vapor will condense into droplets, giving rise to a cloud.

tropopause The inversion that separates the troposphere from the overlying stratosphere. Its altitude varies from approximately 5 mi. at the poles to 11–12 mi. over the equator.

troposphere The lowest region of the atmosphere in which most of the weather and clouds occur. Within it, there is an overall decline in temperature with height.

trough An elongated extension of an area of low pressure, which results in a set of approximately 'V'-shaped isobars, pointing away from the center of the low.

veering A clockwise change in the wind direction, i.e., from East, through South, to West.

wind shear A change in wind direction or strength with a change of position. If, for example, the wind strength increases with increasing height, this is defined as vertical wind shear. If the wind strength changes with motion at a particular level, this is known as horizontal wind shear.

zenith The point on the sky directly above the observer's head.

INDEX

PHOTOGRAPHIC CREDITS

All pictures © Storm Dunlop, apart from those listed below. The Author and the Publishers would like to thank all those who kindly granted permission for their photographs to appear in this book.

t = top, c = center, m = middle, b = bottom

Arnau 143; James S. Barton 89 b; N. Billett 115; Bone 84; Dr. A. Burgueno 8; Trevor Buttress 153; Cinderley 11; Ian Currie 15, 142, 151 c; Robert and Barbara Decker, Double Decker Press 140; Graham Denyer 81 b, 90, 137 c; Ian Dobinson 59 b; L. Draper 146; Karen Dutton 113; Stephen J. Edberg 107, 126, 130, 155 b; J.F.P.Galvin 9, 17 lcm, 26–27, 36, 43 t, 48, 57 c, 66 b, 93, 145; Durham Garbutt 13, 97, 109 t, 112, 149, 164; Dave Gavine 30 b, 83, 85, 103 b, 152; Rosie Holbech 137 t; Isabel Hood 95 b; Dr. Peter Hutchinson 77 b; JDW 31; Anthony Kay 167; R.J. Livesey 172; Lockyer 173; C.A. Matthews 155 t; John McNaught 95 t; Jarmo Moilanen 6, 114, 116, 125, 127, 128; Gordon F. Pedgrift 14; W.S. Pike 17 lcl, 20, 51, 67, 138; Peter C. Roworth 80; Ron Saunders 1, 104, 111, 121, 124; Ian Simpson 89 t, 106, 163; Slebarski 76; Fiona Smith 74; Mike Speakman 79, 175; Stanier 28; Menno van der Haven 108; Vesey 52 t; Waldron 170; Ward 168; Steve Western 81 t; Mike Williams, Akalat Publishing 148 bl; Andrew T. Young 131